UNBIASED STEREOLOGY

UNBIASED STEREOLOGY

A Concise Guide

PETER R. MOUTON

The Johns Hopkins University Press

Baltimore

© 2011 Peter R. Mouton
All rights reserved. Published 2011
Printed in the United States of America on acid-free paper
9 8 7 6 5 4 3 2 1

The Johns Hopkins University Press
2715 North Charles Street
Baltimore, Maryland 21218-4363
www.press.jhu.edu

Library of Congress Cataloging-in-Publication Data

Mouton, Peter R.
 Unbiased stereology : a concise guide / Peter R. Mouton.
    p. cm.
 Includes bibliographical references and index.
 ISBN-13: 978-0-8018-9984-3 (hardcover : alk. paper)
 ISBN-10: 0-8018-9984-2 (hardcover : alk. paper)
 ISBN-13: 978-0-8018-9985-0 (pbk. : alk. paper)
 ISBN-10: 0-8018-9985-0 (pbk. : alk. paper)
 1. Stereology. 2. Microstructure—Measurement. I. Title.
 Q175.M8778 2011
 516.001'13—dc22   2010050256

A catalog record for this book is available from the British Library.

Special discounts are available for bulk purchases of this book. For more information, please contact Special Sales at 410-516-6936 or specialsales@press.jhu.edu.

The Johns Hopkins University Press uses environmentally friendly book materials, including recycled text paper that is composed of at least 30 percent post-consumer waste, whenever possible.

• • •

To D. C. Sterio, whose inspiration, intellect, and resolve to "do it right or don't do it at all" ensured a great deal of scientific progress

CONTENTS

PREFACE

The term *stereology* came to life, so to speak, in May 1961 at a meeting organized by Professor Hans Elias at the The Feldberg, a mountaintop retreat in Germany's Black Forest. The modifier *unbiased* gradually entered the vernacular sometime in the 1980s to differentiate new assumption- and model-free techniques, developed by Hans Jørgen G. Gundersen of Aarhus, Denmark, and colleagues in the International Society for Stereology, from older assumption- and model-based methods in Euclidean geometry, known as *biased* stereology. Since unbiased carries both colloquial and mathematical connotations, *unbiased stereology* has been replaced in many circles by the term *design-based stereology*. Over the past four decades, the theoretical concepts that underlie design-based stereology have been developed, tested, refined, and published by members of the International Society for Stereology under a strong peer-review process in a variety of journals that specialize in stereology, including the *Journal of Microscopy*. Today, the terms *unbiased* and *design-based* refer to the theoretical basis for stereology approaches to estimate first-order parameters, such as number (N), length (L), surface area (S), and volume (V), and several second-order parameters, including variation, spatial distribution, and so on.

Design-based stereology is unbiased in theory, because applications to biological material carry the potential for all manner of assumptions, uncertainties, and artifacts to skew results from expected values. Obtaining unbiased data from design-based stereology is what Luis Cruz-Orive refers to as "a committed task." Every investigator's study design must be firm in identifying and avoiding the stereological bias and uncertainty that arise from a wide variety of tissue-processing and microscopy-related sources. These issues differ considerably from stereology applications in other fields, such as materials sciences, soil sciences, engineering, metallurgy, and geology. The term *organic stereology* is introduced here to highlight biological applications of design-based stereology. Hence, this book is intended to address researchers' problems in the biosciences with applying theoretical principles to biological tissue in an accurate, precise, and efficient manner.

My involvement with stereology began in perhaps the most ordinary way—as a means to an end. I was involved in collaborative research projects in the laboratory of Lars Olson as a predoctoral fellow at the Karolinska Institute in Stockholm, Sweden. One field of study involved transplantation of stem cells into the brains of adult rats. The survival of transplanted stem cells could be demonstrated through staining and microscopy, but progress required reliable morphometric techniques. We needed a morphometric approach to determine whether certain molecules delayed or enhanced the growth of these undifferentiated cells. I had some experience using "biased stereology," that is, clicking with a hand-counter to count hundreds of cell profiles on dozens of thin tissue sections at low magnification—one click for each cell profile. Following the application of a so-called correction factor, this approach generates biased data for the number of cells per unit area (areal density). Besides being tedious, this approach is also labor and cost intensive, and based on numerous faulty, model-based assumptions (e.g., assume the cell is a sphere).

Hoping for a more accurate and efficient approach, I waded into the stacks of journals at Den Karolinska Bibliotek, the revered library at the Karolinska Institute. This research turned up various morphometric approaches for sampling and counting cells, with the most highly cited papers reporting data based on thousands and even hundreds of thousands of clicks on cell profiles. One study that took years to finish reported "complete counts" based on clicks of more than 400,000 cell profiles on literally hundreds of tissue sections. Notwithstanding the tremendous work and dedication of this effort, this approach suffered from the same weaknesses as my own. The counting method assumed that the neurons of interest were spheres of uniform size, shape, and orientation, an assumption violated by not only the cells in that study but also all cells in the fields of bioscience in general and neuroscience in particular. Despite the faulty assumptions, models, and correction factors in such studies, these approaches were highly regarded for their enormous investment of material and personnel resources. Those were the days before books on unbiased stereology and commercially available computerized stereology systems. I found a recently published article in the *Journal of Microscopy* by D. C. Sterio entitled, "The Unbiased Estimation of Number and Sizes of Arbitrary Particles Using the Disector." This theoretical work proposed a method for counting the number of cells on tissue sections, without any further assumptions about the size, shape, or other characteristics of the cells. To my novice eye, this seemed like the way to go.

After listening to the outcome of my literature survey, Lars Olson suggested that I attend a stereology course offered by Hans Jørgen Gundersen at the nearby University of Aarhus in Denmark. I met an international team of dedicated stereologists from a wide variety of scientific disciplines, including Mark J. West, Bente Pakkenberg, Eva Jensen, and Arne Møller (Denmark), Vivian Howard (Great Britain), Luis Cruz-Orive (Spain), Terry Mayhew (Scotland), and Adrian Baddeley (Australia), all dedicated to helping bioscientists quantify organic tissue. Although I never met Dr. Sterio at that meeting (or did I?*), I learned that standard histological sections were biased for number; that is, accurate counts of the number of neurons could never be obtained from them, no matter how many thousands of clicks on cell profiles were carried out; that *greater* numbers of clicks on cell profiles actually produced *less* accurate data; and that the number of surviving stem cells in the brains of my rats could be accurately quantified using the disector method. I also learned that Gundersen's students nicknamed him "Indiana Jones of Anatomy" for his ubiquitous leather jacket, unrelenting travel to far-flung places, and fervent dedication to design-based approaches in stereology.

After this course, I accepted Gundersen's invitation to join his research group as a postdoctoral fellow after completing my PhD defense. After two years of studying stereology principles and practices

---

* D. C. Sterio is the nom de plume of a stereologist who does not wish his or her name to be associated with the disector method.

and living the Danish way of life, I received an overseas phone call from Donald L. Price, head of the Neuropathology Laboratory at the Johns Hopkins University School of Medicine in Baltimore, Maryland. Dr. Price told me that, while he did not fully understand stereology, "everyone is telling me that it's important." I accepted his invitation to join his research group at Johns Hopkins as a National Institutes of Health Fellow in experimental neuropathology, where I spent much of the next two years informing U.S. bioscientists about the principles and practices of stereology. I typically faced skepticism, resistance, and aggressive questioning about why my stereology was better than the assumption- and model-based methods used for decades to quantify biological structure.

After the neuropathology fellowship, I joined the faculty in the Department of Pathology at Johns Hopkins and spent the next six years working to promote applications of design-based stereology to the biological sciences. Together with Arun M. Gokhale (Georgia Tech, Atlanta) and Mark J. West (University of Aarhus, Denmark), I started a long-running training program, Applications of Unbiased Stereology to Neural Systems, to train bioscientists in the theory and practice of state-of-the-art design-based stereology. To increase the throughput of research projects, we incorporated design-based stereology into integrated hardware-software computerized systems with support from the Johns Hopkins University School of Medicine and in collaboration with Joel Durgavich at Systems Planning and Analysis in Alexandria, Virginia. To help disseminate stereology-related information to the international biomedical research community, in 2000, we established the Stereology Resource Center (SRC) with a singular mission, "to bring state-of-the-art stereology to the biosciences."

Among the important changes in stereology during the past two decades, I have identified the following three as perhaps the most important. First, concerns about design-based stereology have shifted from opposition ("Why is design-based stereology better than my current methods?") to support ("Since design-based stereology can produce reliable results, why accept the results of biased approaches?"). Second, efforts from members of the International Society for Stereology to promote unbiased approaches have attracted the attention of editors and reviewers of bioscience journals, as well as federal and private funding agencies in many countries, which, in turn, have changed the priorities for funding and publishing bioscience research. Third, and perhaps most significant, I have seen the application of design-based stereology to organic material, that is, organic stereology, become the dominant method for quantifying the biological structure in organic tissue.

This book celebrates the 50-year anniversary of the first stereology meeting in the Black Forest of Germany. The opinions expressed here are my own. I wish to express my appreciation to Vincent J. Burke and the editorial and production staff at the Johns Hopkins University Press, and to my wife, Sammie, for her inspired dedication to prepare images and continued support throughout the long ordeal of processing words into sentences, sentences into paragraphs, and so on. I welcome positive and negative feedback from readers whom I trust will appreciate this text as I would have as a graduate student many years ago.

UNBIASED STEREOLOGY

1 ...

ELIAS COINS A WORD

In the years leading up to 1961, an interesting point occurred to German histologist Hans Elias. He realized that researchers in a wide variety of scientific disciplines, including geology, materials sciences, engineering, and natural sciences, struggled with a ubiquitous problem: how to quantify structural parameters of three-dimensional objects based on their two-dimensional representations on cut sections.

Several innovations in the late 1950s and early 1960s provided the stimulus for biologists and histologists to acquire reliable methods to quantify biological objects in tissue sections. Antibody-based probes to label specific populations of biological objects based on the presence of specific proteins (antigens), a technique called immunocytochemistry, arose in the 1960s from the research of A. H. Coons. This approach spurred the development of other immunohistochemical techniques to quantify the amount of biological substances in tissue. Together, these histological techniques allowed histologists to distinguish, for the first time, new and discrete populations of cells, fibers, and blood vessels from one another, leading to questions about relative quantities and changes in these structures during disease and aging and following experimental treatments.

In concert with these methodological breakthroughs in histology, the availability of affordable, high-quality, mass-produced glass and electromagnetic lenses in the late 1950 and early 1960s provided biologists with their first glimpses of previously unknown biological structure. Researchers in the biosciences used high-resolution light and electron microscopy to address a variety of lingering controversies about tissue structure at the cellular and subcellular levels. These insights led to surprising answers, followed by new questions and hypotheses for testing. However, the primary approaches to quantify tissue at that time relied on experts in different disciplines for subjective, nonparametric assessment of biological tissue that generally lacked a strong mathematical basis and suffered from poor inter-rater reliability. Finally, based on groundbreaking work of theoretical statisticians leading up to the 1960s, including W. Edwards Deming (United States), Ronald Fisher (United Kingdom), and C. R. Rao (Canada), bioscientists had powerful methods to test hypotheses using inferential statistics, but these approaches required parametric data. Rather than traditional expert-based qualitative analysis, bioscientists needed hard data to input into these new statistical approaches. For progress to continue, bioscientists increasingly recognized the requirement for more sophisticated morphometric methods to quantify organic structure.

For example, they needed to know what volume, length, and surface area of biological structure were present under normal conditions? How many cells were lost during aging, and what was the effect of disease? How did these parameters vary under the influence of various toxins? Together with behavioral and physiological endpoints, these morphological data could provide critical insight into structure-function relationships for a variety of tissues and organisms.

Histologists such as Elias shared a strong interest in learning methods developed by other disciplines to quantify 3-D objects based on their appearance on 2-D sections. Almost daily, histologists then and now deal with this issue in their work, trying to understand changes in 3-D objects in biological tissue based on 2-D representations of those objects on tissue sections. Two-dimensional microscopic images do not represent the full dimensionality of 3-D objects as they exist in unsectioned and unstained tissue blocks and especially not in the living organism. Yet how does one make these extrapolations from the appearance of organic objects on 2-D sections to their quantities in 3-D tissue? In the early 1960s, Elias looked to other fields for inspiration and support.

Since around the start of the twentieth century, geologists in the commercial sector had been using a series of ad hoc approaches to estimate the volume of petroleum in underground oil fields, based on core samples. The samples were cut (sectioned) with saw blades, and measurements were obtained using probes placed at random over the cut surfaces of the resulting sections. However, the primary interest lay not in the amount of visible oil on the cut surfaces but rather on the 3-D deposits of oil in the ground, that is, the source of the core samples. Considering the enormous time, expense, and effort expended to drill in a particular area, these geologists required reliable methods to estimate the volume of oil in the ground. That is, they needed a function for converting area estimates from the oil phase on the cut surfaces of core samples into volume estimates for the oil deposits in the ground. Materials scientists and engineers struggled with their own version of the same problem. During the development of turbine engines for jet airplanes, for example, imaging studies revealed that small pockets of water in turbine blades could cause blades to fracture, resulting in engine failure during flight. Scientists could observe and measure the area of water pockets on 2-D sections through the turbine blade but needed to know the volume of water associated with these areas.

In his survey of other disciplines, Elias discovered that scientists in some fields had developed solutions to this problem, some of which could help scientists in other disciplines. These solutions do not easily disseminate across disciplines because researchers in one field are usually unaware of work accomplished in other fields. Scientists keep up with developments in their respective fields by attending conferences and reading their peer-reviewed literature. However, in general, geologists do not attend conferences or follow the peer-reviewed literature of biologists and vice versa.

To address this large and growing demand, Elias organized a meeting for May 11, 1961, at a mountaintop resort called The Feldberg in the Black Forest of Germany, and invited experts from various scientific disciplines to attend. During the conference, Elias proposed the name *stereology*, the study of 3-D objects, from the Greek *stereos* (στερεος = solid). Stereology, it was decided, would refer to the quantitative analysis of 3-D objects based on their 2-D appearance on cut sections. After this historic meeting on The Feldberg, a short note appeared in the journal *Science*, along with a description of the proceedings and a list of attendees. In the months following this first stereology meeting, Elias's mailbox flooded with requests from geologists, engineers, biologists, and materials scientists who had not attended the meeting on The Feldberg but, nevertheless, wanted to receive more information and plans for similar events in the future; they, too, had problems to address with stereology. Sensing the proverbial tip of an iceberg, Elias organized the International Stereology Congress in Vienna, Austria, on April 18–20, 1963. At that conference, the attendees voted to establish the International Society for Stereology, with Professor Hans Elias elected to serve as the founding president.

2 . . .

SOLID 3-D OBJECTS

- The World of Euclidean and Non-Euclidean Objects
- Biological Structures as Naturally Occurring, Organic Structures
- Definition of Organic Stereology

Before we delve into the world of stereology, the quantitative study of 3-D objects, here are a few words about 3-D objects in general. Unless you are reading this book in a forest, looking out to sea, or wandering in a desert, then you are likely surrounded by a variety of man-made objects. Methods to quantify man-made objects (stereometry) dates to the first scientists, the ancient Egyptians, but these methods underwent a major advance ca. 300 BC with the publication of Euclid's *Elements*, a 13-volume compilation of the state of geometry. Considered one the greatest books of all time, Euclid's series outlined the formulas, theorems, and proofs to quantify essentially all types of classically shaped objects. For the next 23 centuries, architects, engineers, designers, and manufacturers across the planet relied on the concepts in Euclid's *Elements* to construct architectural elements, furniture designs, and many other objects built by human hands or machines, with heavy reliance on planes, spheres, lines, cylinders, rectangles, and topological elements. The study of organic stereology requires an appreciation of the fundamental distinctions between objects with non-Euclidean and Euclidean shapes.

As shown in fig. 2.1, essentially all 3-D objects fall into the categories of either naturally occurring or man-made. Man-made objects generally conform to Euclidean shapes and their derivatives. These

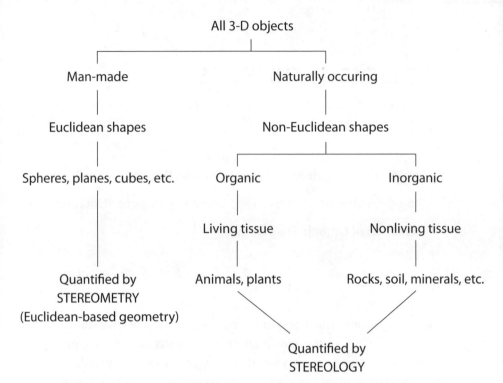

FIG. 2.1. **Classification of 3-D Objects.** Classification scheme for objects quantified by stereology and stereometry.

forms appear in building architecture, furniture, manufacturing, textiles, and a myriad of other products and textiles. The other category of 3-D objects includes the full array of organic (living) and nonorganic (nonliving) objects.

The modern term for quantitative analysis of classically shaped Euclidean objects is *stereometry* (formerly solid geometry). Naturally occurring 3-D objects generally exist in non-Euclidean shapes. This classification includes the organic (once-living, carbon-based) single-cell prokaryotes (bacteria) and all multicellular plants and animal tissues and the inorganic (nonliving) structures, such as rocks, minerals, clouds, and soil. In the biosciences, we primarily focus on the organic objects that exist in the natural world and the somewhat unnatural world of the research laboratory. For organic objects more complex than the simplest prokaryotes, biological variability produces a range of sizes and nonclassical shapes that violate the models required for applying stereometry to quantify structural parameters such as volume, surface area, length, and number. Despite their similar names, striking differences exist in applying stereology and stereometry to biological objects, as illustrated by a well-known vignette involving the Greek philosopher-mathematician Archimedes (287–212 BC) and the principle of volume determination by water displacement.

During the Renaissance of the fifteenth and sixteenth centuries, western Europeans translated the ancient writings of Greek scientists, philosophers, and artists. *De architectura libri decem* (*The Ten Books on Architecture*), by the Roman architect Vitruvius in the first century BC, tells the story of Hiero, who ruled the city-state Siracusa in ancient Greece during the third century BC. Around 260 BC, Hiero decided to commission a gold crown as an offering to the gods in return for a recent period of prosperity. The king then provided a quantity of gold to the royal goldsmith, who proceeded to create a delicate gold crown in the natural form of smoothly curving vines connecting a series of variably shaped leaves and twigs. Shortly thereafter, the king became suspicious that the crown was not pure gold but rather contaminated with silver and, therefore, neither to the standard of his instructions nor of sufficient quality to offer the gods. Although the king had a number of destructive methods at his disposal to test his hypothesis, for example, melting the crown into various components, he preferred not to risk damaging the crown and possibly angering the gods, in the event his suspicions about the royal goldsmith were unsubstantiated. Finding himself at an impasse, Hiero requested help from Archimedes, who, though still in his early twenties, was already an accomplished mathematician and engineer. Archimedes quickly arrived at the crux of the problem. Since the density (mass per unit volume) of pure gold was known, and the weight of the crown could be determined with high accuracy, he needed a method to determine the volume of the crown. Euclid's classical work on the geometry of solids, *Elements*, had been published less than a half-century earlier (ca. 300 BC) and was well known to Archimedes; yet it held no formulas to quantify the volume of a non–classically shaped object.

According to Vitruvius' account, Archimedes pondered the question for some time before realizing the answer while lowering himself into a bathtub—the amount of water displaced by the crown must equal the volume of the crown, without any assumptions about the crown's three-dimensional shape. In effect, Archimedes developed the first method to quantify the volume of an arbitrary (nonclassically) shaped 3-D object.

Before the 1960s and the era of modern stereology, many biologists sought to quantify biological objects using Euclidean formulas, approaches that required a range of nonverifiable assumptions, models, and correction factors. Like Hiero's crown, however, organic objects rarely, if ever, conform to the classical shapes required by Euclidean formulas. In a sense, these methods tried to "force" organic structures into Euclidean formulas. The rejection of Euclidean-based approaches for quantifying non-Euclidean objects led to the modern field of unbiased stereology. Because it relies on study design, rather than on Euclidean models, unbiased stereology is often referred to as *design-based* stereology. Since the term *stereology* may also refer to quantification of, for example, nonorganic objects in geology and the materials sciences, the term *organic stereology* refers to the application of design-based (unbiased) stereology to organic structures.

3 . . .

REGIONAL VOLUME ESTIMATION

- Theoretical Basis for the Cavalieri Theory of Indivisibles
- Contrast Reference Space with Region of Interest (ROI)
- Use of the Cavalieri's Principle to Estimate V_{ref}

The Theory of Indivisibles

Beginning in the 1960s, biologists in the International Society for Stereology learned about the principles and practices of stereology that had been used for centuries by members from the fields of materials sciences and geology. One of the first concepts they encountered was named after the Italian mathematician Buonoventura Francesco Cavalieri (1598–1647), a student of Galileo Galilei. Under the guidance of his mentor, Cavalieri carried out scientific research in mathematics, optics, mechanics, and astronomy during the first half of the seventeenth century.

Years earlier, at age 22, Galileo had published his first book, *La bilancetta* (*The Little Balance*), which detailed a practical device for accurately and precisely determining the purity of gold and other precious metals based on the Archimedes principle. Notwithstanding the value of this device in the world of commerce, Galileo's balance suffered from the same limitations as the Archimedes principle: the approach could neither be applied to porous material nor to finding the volume of a smaller object embedded within a larger one. This state

9

of affairs led Cavalieri to research methods for volume quantification of arbitrary-shaped objects that could be cut into sections, an effort that led Cavalieri to publish a scientific paper, "Geometria indivisibilibus continuorum nova quadam ratione promota" ("Geometry Developed by a New Method through the Indivisibles of the Continua") in 1635.

In addition to overcoming the limitations of the Archimedes principle and Galileo's little balance, the Cavalieri principle established an unbiased method to estimate the regional volumes of organic objects, based on uniform-random sections through the objects. Specifically, Cavalieri's theory of indivisibles states that any volume of matter is contained within an infinite number of parallel planes, which, when stacked together, form the volume in question. As shown in fig. 3.1, the reference volume, V_{ref}, contained within a stack of coins can be determined from both the Euclidean formula for the volume of a cylinder (product of radius and height), regarding the coins on the left, and from the product of the sum of the areas of each coin (ΣA) and T, the distance between the face of each coin.

$$V_{ref} = \Sigma A \cdot T \tag{3.1}$$

In contrast to Euclidean approach, the theory of indivisibles carries no requirement about the shape of the cylinder, only that the planes (coins) are parallel. Cavalieri's theory further states that any planar area contains an infinite number of parallel lines, which when

FIG. 3.1. Cavalieri's Theory of Indivisibles. Volume estimation for classical shape (*left*) and arbitrary shape (*right*).

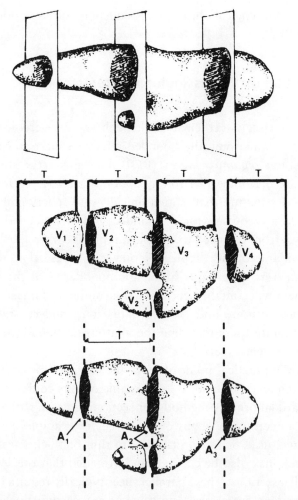

FIG. 3.2. Cavalieri Principle for Arbitrary Objects. First cut with a random start, systematic uniform random thereafter at a distance, T. The total V is the product of T and the sum of areas on each cut surface (total $V = T \cdot \Sigma A$).

arrayed side by side, form the area; and every line contains an infinite number of adjacent points that, when arrayed side-by-side form the line. A century later, both Isaac Newton and Gottfried Leibniz relied on Cavalieri's theory of indivisibles in their postulations of infinitesimal calculus.

In current applications of organic stereology, the Cavalieri principle provides an accurate, precise, and efficient method for volume estimation of biological objects, independent of all assumptions about shape (fig. 3.2).

Reference Space versus Region of Interest

Accurate, precise, and efficient results with organic stereology require a well-defined, anatomically bounded reference space in each study. In contrast to a *region of interest* (ROI), a term from the image analysis literature that may refer to any region or area, a *reference space* represents a structure that is anatomically and functionally defined. The simple prerequisite for a bounded and well-defined reference space arises from the need to sample a reproducible region of tissue. If the boundaries of the sampled region vary because of nonbiological sources, such as uncertainty about the boundaries of the structure, then the data will contain a corresponding amount of uncertainty. For this reason, accurate and reliable data from modern stereological methods require users to define the anatomical boundaries of their particular reference space.

From the practical point of view, the Cavalieri principle to estimate the reference volume, V_{ref}, of organic objects conforms easily to the standard laboratory methods for sectioning tissue with a knife attached to a microtome, or any other device, for example, a cryostat or a vibratome, used to section tissue into thin or thick parallel slices. The volume may also be cut freehand provided that the V_{ref} is sliced into parallel sections. These parallel slices typically require between 8 and 14 sections cut as serial sections at any section thickness, also known as the *block advance* (typically 10 to 50 μm). If the V_{ref} is contained within fewer than 14 sections, then the volume estimate is carried out on all of the sections, that is, without subsampling. If the volume is contained within 15 or more sections, then the most efficient approach is a subsample from 10 sections throughout the volume of interest. This subsampling is done by calculating a section sampling fraction (ssf) that will result in about 8–12 sections. For example, if 110 planes of sections are obtained through the volume of interest, sampling every 11th section (ssf = 1/11), with a random start in the interval 1–11 will result in 10 sections. If the volume for the next subject is contained by greater or lesser numbers of planes, the same approach is applied to obtain about 10 sections (range 8–12) through the V_{ref}; that is, for a V_{ref} contained within 90 sections, sampling every 9th section (ssf = 1/9) with a random start in the interval 1–9 will result in 10 sections. Subsampling fewer than about 6 to 8 sections through a volume of interest increases the chance of not capturing the between-section variation in the object, while subsampling more than 16 sections through an object simply increases the workload without improving the volume estimate. Thus, for volume estimates of essentially any group of biological objects, the ideal number of sections to analyze per object falls between about 8 and 14 sections.

Although organic stereology is carried out on 2-D sections through 3-D objects, it helps to maintain a view of biological objects in 3-D. For instance, a particular V_{ref} is sliced along a single axis, for example, coronal; the number of resulting sections that contain the V_{ref} is not directly related to the *magnitude* of the V_{ref}. Because there is a limited selective pressure on biological objects to conform to the exact size and shape, their sizes and shapes in 3-D are free to vary across individuals in the same group. For instance, a V_{ref} contained by 110 sections may be equivalent in volume to a V_{ref} contained by 90 sections. Naturally occurring, organic structures from a similarly treated group are not like microscopic radios manufactured using a common blueprint, with each transistor located an exact distance from each capacitor. Anatomically distinct regions of biological tissue do not exist as cylinders of equal diameter but rather as objects with variable areas on each section. Thus, a V_{ref} for one individual may be tall in the dorsoventral axis and short in the anterioposterior axis and could have similar volume to another individual's V_{ref}, that is, shaped short in the dorsoventral axis and tall in the anterioposterior axis.

The Cavalieri principle integrates the reference areas across the total number of sections to estimate the total reference volume, V_{ref}. For the estimate of V_{ref} to be unbiased (accurate), two prerequisites must be met: (1) the first cut through each reference volume must be at random and (2) the planes cut through the volume must be parallel. Selecting the first section at random in the sampling interval ensures that all areas through the V_{ref} have an equal chance of being sampled, which, in turn, guarantees the average estimate of V_{ref} for all individuals in a particular group, for example, control or treatment, will estimate the expected (true) values of V_{ref} and biological variability (BV) of the parameter for the population. In the mathematical sense, an estimate of V_{ref} and BV is unbiased when increased sampling causes the estimate to progressively converge on the true values for the population.

In later chapters, we will discuss BV, together with the other factor that contributes to the variability observed in organic stereology—variation due to sampling (error variance). Error variance arises from sampling variability at two levels within each individual, between-section error and within-section error. Between-section error, also known as the variation from systematic-uniform-random (SUR) sectioning, refers to variability in V_{ref} estimates that arise from different stringencies of sampling through the reference space, for example, using 5–6 sections versus 10–12 sections to estimate V_{ref}. The second factor that contributes to error variance is variation that arises from the number of probes placed within sections. In chapter 13, we discuss how two factors, the average coefficient of error and BV for a given group, are used to optimize the amount of work required to optimize all sampling designs to achieve maximal efficiency, a concept known as "Do More, Less Well."

4 . . .

AREA ESTIMATION
BY POINT COUNTING

- Delesse Principle to Relate Area Fraction to Volume Fraction
  ($A_A = V_V$)
- Point Counting Stereology for Unbiased and Efficient Area
  Estimation
- Advantages of Total Volume (Load) over Ratio Estimators for
  Biological Tissue

In the early 1960s, biologists became acquainted with many of the stereology principles other members of the International Society for Stereology used to quantify solid objects. Among these concepts was the Delesse principle, an unbiased method to estimate the 3-D volume fraction of an object based on the corresponding 2-D area fraction on cut surfaces through the object. August Delesse was a French geologist who first reasoned that 2-D sections through 3-D core samples might be used to make accurate predictions about the quantity of oil deposits contained in the core samples.

Starting from the observation that the 3-D volume of oil deposits, V_{oil}, in a core sample appeared to correlate to the reference area, A_{oil}—the oil phase on a random cut surface through the core sample—Delesse devised a simple yet elegant experiment to test this hypothesis (Delesse, 1847). He collected core samples from an oil field and sawed through these samples to reveal the A_{oil} on each of the cut surfaces. To quantify the A_{oil} from the oil phases on these cut surfaces, using tracing paper, he outlined the oil phase on the cut surfaces. Next, using scissors, he cut out these outlined areas and weighed them on a scale; the weight of these cut-out regions provided an accurate correlate to their A_{oil} on the cut surfaces. To determine the volume fraction of oil in the core sample, he first determined the volume of the rock, V_{rock}, pulverized the core samples, and then used chemical methods to measure the volume of oil, V_{oil}. Through correlations across a range of core samples, Delesse showed that the area fraction, that is, the ratio of the area of the oil phase on the cut surface (A_{oil}) to the area of the cut surface of rock (A_{rock}), was directly related to the ratio of the volume of oil in the rock (V_{oil}) to the volume of the rock (V_{rock}), also known as the Delesse principle, as shown in equation (4.1).

$$A_{oil}/A_{rock} = V_{oil}/V_{rock} \tag{4.1}$$

In 1847, Delesse presented his observations to the Proceedings of the Royal Academy of Sciences in Paris in a paper entitled "Procédé mécanique pour déterminer la composition des roches" ("A Mechanical Procedure to Determine the Composition of Rocks").

Throughout the second half of the nineteenth century, geologists used the Delesse principle as a basis for geologic explorations. In 1898, another geologist, A. Rosiwal, showed that the same results could be obtained by randomly tossing an array of lines (line-probe) over the cut surface of the rock and then making two measurements: (1) the lengths of line over the oil phase (L_{oil}) and (2) the length of lines crossing the cut surface of rock (L_{rock}). The ratio of these two area estimates constitutes the line fraction, that is, L_{oil}/L_{rock}. By repeating the experiment in Delesse's paper, Rosiwal showed in 1898 that the line fraction of oil on random cut surfaces is equivalent to the volume of oil in the rock samples (eq. 4.2).

$$L_{oil}/L_{rock} = A_{oil}/A_{rock} = V_{oil}/V_{rock} \tag{4.2}$$

By substituting a hand ruler for measuring L_{oil}/L_{rock} and eliminating the more labor-intensive tracing/cutting/weighing process proposed by Delesse, Rosival considerably improved the efficiency of this method. In 1933, the geologist A. A. Glagolev introduced a third simplification, the point counting method. Rather than using a grid with a known length of lines to probe the oil phase on the cut surface,

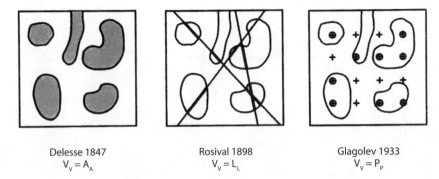

FIG. 4.1. **Delesse Principle.** Historical development of Delesse principle from Delesse ($V_V = A_A$) to Rosival ($V_V = L_L$) to Glagolev ($V_V = P_P$).

Glagolev used a grid of points (+) with a known area per point, $a(p)$, corrected for areal magnification $[a(p)/mag^2]$. The $a(p)$ is the product of the distance between points in the x-axis and the distance between points in the y-axis. In agreement with the methods proposed by Delesse and Rosival, the work of Glagolev and Thomson showed that P_{oil}/P_{rock}, the point fraction of oil on a random cut surface, is equivalent to V_{oil}/V_{rock}. Combining these methods leads to the following equations based on the Delesse principle (fig. 4.1; eq. 4.3):

$$P_{oil}/P_{rock} = L_{oil}/L_{rock} = A_{oil}/A_{rock} = V_{oil}/V_{rock} \qquad (4.3)$$

H. W. Chalkley (1943) was the first biologist to use point counting stereology and a wide range of applications followed on biological objects, including cells, extracellular deposits, and pathological features (fig. 4.2).

Careful reflection on Fig. 4.3 reveals that the size of the point (+) used to probe the area of interest could introduce bias—larger points would be expected to intersect a greater fraction of the area of interest than smaller points. To avoid bias related to the size of the point, point counting is done by preselecting a dimensionless "point on the point" (see arrow) to use as the consistent counting item. After random placement of the grid over the reference area, only size of the reference space, that is, the reference area, affects whether a point-object intersection occurs.

These studies require randomly placing a point grid with known

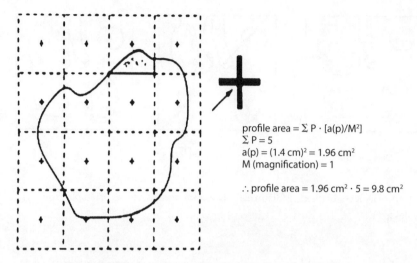

profile area = Σ P · [a(p)/M²]
Σ P = 5
a(p) = (1.4 cm)² = 1.96 cm²
M (magnification) = 1

∴ profile area = 1.96 cm² · 5 = 9.8 cm²

FIG. 4.2.   Point Counting Example. Area estimation using a randomly placed point grid. The sum of points (ΣP) counted when dimensionless point (indicated by arrow) on point grid with area per point $a(p)$ adjusted for area magnification intersects the profile of interest.

$a(p)$ over the cut surfaces of tissue sections 8–12 through the V_{ref} stained to show the biological objects of interest. The data collection involves counting the number of points that intersect the biological object of interest, P_{object}, and the number of points that intersect the reference area, P_{ref}. The optimal level of sampling stringency for point counting is achieved by setting the $a(p)$ to ensure that between 100 and 200 points intersect the area of interest across all 8–12 sections through the V_{ref}. Taking the sum of points that hit objects of interest and the reference area provides an unbiased estimate of the point fraction, $\Sigma P_{object}/\Sigma P_{ref}$. Finally, using the $a(p)$ for the point grid to convert points to areas, that is, ΣP to ΣA, allows for estimating area fraction for the biological objects within the V_{ref}.

$$\text{area fraction} = [\Sigma P_{object} \cdot a(p)]/[\Sigma P_{ref} \cdot a(p)] = \Sigma A_{object}/\Sigma A_{ref} \quad (4.4)$$

According to the Delesse principle, the area fraction is directly proportional to the volume fraction, with no further assumptions:

$$\Sigma A_{object}/\Sigma A_{ref} = \Sigma V_{object}/\Sigma V_{ref} \quad (4.5)$$

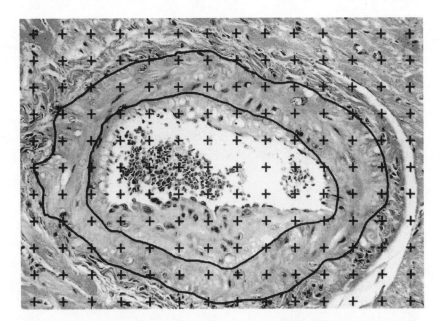

FIG. 4.3. Point Counting Aorta. Example of point counting to estimate of two reference spaces in rat aorta: volume of vascular pathology (*neointima*) within lumen; and volume of *tunica media* within dark drawn lines. Areas estimated by point counting entered into Cavalieri principle to estimate total *V* for each reference space.

In a practical sense, the Delesse principle allowed for estimating areas on the cut surfaces of sections, as required an unbiased estimate of an anatomically defined reference volume, V_{ref}, using the Cavalieri principle (Gundersen and Jensen, 1987).

Figure 4.3 illustrates an application of point counting stereology to estimate the volume of pathology within the lumen of the aorta of a heart undergoing rejection following transplantation (Luo et al., 2004). As the host tissue reacts to the foreign tissue, macrophages move into the lumen, leading to progressive narrowing of the vessel and eventually rejection of the transplanted heart. The progression of this rejection process can be observed by removing the aorta at various periods after transplantation and cutting serial coronal sections through the long axis of the aorta. The Cavalieri point counting method then allows us to estimate the total volume of neointimal pathology as a function of time after transplantation.

Thus, the point counting method provides a straightforward, accurate, and efficient approach to estimate a biological area of interest, A_{ref}, as the product of the area per point $[a(p)]$ for the point grid and ΣP, the number of point-area intersections. As this example demonstrates, it is not necessary to *measure* area, for example, with a planimeter, a relatively time- and labor-intensive technique that requires tracing the area of interest. Instead, the more effective approach is "sample don't measure," a central tenet in organic stereology.

The Reference Trap

At this point, we arrive at a significant difference between the applications of design-based principles to organic versus inorganic objects. In contrast to cutting nonorganic materials, such as rocks and mineral-based polymers, which is usually carried out using heavy saws and other cutting tools, biological objects are visualized using protocols that involve dyes and a variety of dehydrating agents, such as heat, alcohols, and aldehydes, which cause changes in the 3-D structure of the tissue. That is, these treatments enhance the visualization of biological objects but may also cause massive tissue shrinkage, typically reducing tissue volume by 15% to 50% or more, depending on the particular tissue-processing protocols. If this dehydration were a consistent process across tissues, there would be no cause for concern; however, numerous studies (Dam, 1979; Boonstra et al., 1983; Haug et al., 1984; Gardella et al., 2003) have reported significant variations in tissue shrinkage for the same tissues from different species (human vs. rodent) and for tissue from the same species that undergo different treatments, for example, young versus aged, lesioned versus control, healthy versus disease. This potential for differential tissue shrinkage between groups could confound results and conclusions based on ratios such as area fraction. By making inferences about the effects of experimental treatments, this nonuniform shrinkage of tissue could confound conclusions based on the Delesse principle. The assumption that ratio estimators such as volume fraction provide an equal substitute for estimates of total parameters such as volume carries the risk of introducing a well-known source of stereological bias known as the *reference trap* (Braendgaard and Gundersen, 1986).

Fortunately, there is a simple method that avoids this potential source of error in most cases. Since shrinkage occurs to an equivalent extent in both V_{ref} and the denominator of V_{object}/V_{ref}, any shrinkage that occurs in the tissue cancels in the product of V_{object}/V_{ref} and V_{ref}, leading to an unbiased estimate of the total volume of the biological object of interest, total V_{object}.

$$\text{Total } V_{object} = (V_{object}/V_{ref}) \cdot V_{ref} \qquad (4.6)$$

The same approach leads to calculating total distance traveled as the product of speed (miles per hour) and time spent traveling:

$$\text{total distance (miles)} = \text{miles/hour} \cdot \text{traveling time (hours)}$$
$$500 \text{ miles} = 50 \text{ miles per hour} \cdot 10 \text{ hours} \qquad (4.7)$$

In this example, the units of hours cancel in the product, which leaves miles as the correct unit.

Calculation of total V_{object} in equation (4.7) is an effective means to avoid error arising from differential tissue shrinkage in all cases, provided that volume fraction (V_{object}/V_{ref}) and the reference volume (V_{ref}) are estimated on tissue sections following identical tissue processing. For example, equation (4.7) would not avoid error due to differential tissue shrinkage if the two multipliers, V_{object}/V_{ref} and V_{ref}, are estimated on stained and unstained tissue, respectively. Finally, in some studies, especially those involving biopsy tissue, the entire reference volume may not be available for estimating V_{ref}. These cases require caution to avoid falling into the reference trap through interpretations and conclusions based on results from ratio estimators, rather than total parameters.

5 . . .

PROBE-OBJECT INTERSECTIONS

- **Point Counting Represents Probability Theory Applied to Stereology**
- **Unbiased Counting Rules for Point-Area Intersections**
- **Selection of a Dimensionless Point**

Unbiased stereology uses probes with known geometric properties to estimate the four first-order parameters, volume (V), surface area (S), length (L), and number (N). The geometric probes used to estimate these parameters share an interesting relationship vis-à-vis their respective dimensions. As shown in table 5.1, for a parameter-probe combination to be unbiased, the sum of dimensions in the probe and the parameter must equal at least 3, since there are a total of three dimensions in tissue.

Unbiased stereological estimators use geometric probes to estimate one of the four first-order stereological parameters, V, S, L, N, as well as second-order parameters, such as spatial distribution, variation, and two-point correlations. Regarding the point counting method, to estimate an anatomically defined reference area, A_{ref}, we use an approach that avoids all Euclidean-based assumptions and models. Using a randomly placed point grid, the estimate of A_{ref} varies directly with the expected (true) reference area in question. When this requirement is met, the result is unbiased (accurate). Consider the example in fig. 5.1.

Figure 5.1 shows a reference area of interest, A_{ref}, indicated by a hatched area, α. According to classical probability theory, if point z with area per point, a, is thrown at random over α, the formula (P ·

TABLE 5.1. **Parameter-probe dimensions**

Parameter	Dimension	Structure	Probe	Dimension	Sum of Dimensions
Volume	3	Volume	Point	0	3
Area	2	Surface	Line	1	3
Length	1	Curve	Plane	2	3
Number	0	Cardinality	Disector	3	3

a) provides an unbiased α estimate, as indicated by (α). P is the probability of an intersection between point z and α.

Outcome 1: $P = 1$ if point z intersects α.

Outcome 2: $P = 0$ if otherwise.

Using the identities in table 5.2, summing the probabilities for the only two possible outcomes, either z hits the hatched region ($P = 1$) or not ($P = 2$), we can show that an unbiased estimate of α, indicated by (α) and calculated using the formula (α) = $a \cdot P$, is equal to the true value of α:

$$(\alpha) = [(\alpha/a) \cdot a] + [(1 - \alpha/a \cdot a) \cdot 0]$$
$$(\alpha) = \alpha + 0$$
$$(\alpha) = \alpha$$

Thus, est (α) = $a \cdot P$ is an unbiased estimator of α.

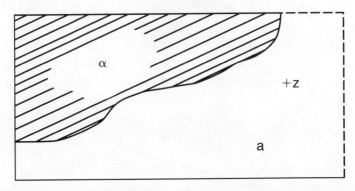

FIG. 5.1. **Probe Intersection Probability.** When point (z) with area per point, a, is placed at random over profile area of interest, α, there are only two possibilities: Z intersects α ($P = 1$); or Z does not intersect α ($P \neq 0$).

TABLE 5.2. Probabilities for α estimation by (α) = a · P

P	Pr(P)	(α) = a · P
1	α/a	a
0	1 − α/a	0

To avoid ambiguity about points that fall exactly on the boundary of the hatched region, we follow the convention to include the reference area within the area outlined. This way, all points that fall exactly on the boundary will always be counted. The following section discusses the need for one final counting convention, an approach that avoids introducing bias by using points of different sizes.

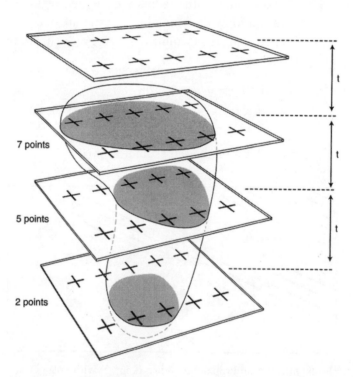

FIG. 5.2. Cavalieri Point Counting in 3-D. Points intersect reference areas (*shaded*) on planes separated by distance, *T*.

Unbiased Rule for Counting Point-Area Intersections

For a stereological method to be theoretically unbiased, the probe must not introduce bias into the estimation procedure. With point counting, this requires a consistent rule to resolve uncertainties about what constitutes an intersection between each point on the point grid and the boundaries of the reference area; that is, when does point z intersect the hatched region in fig. 5.2? Counting all points without regard to the area covered by the actual point would introduce bias (systematic error). That is, points on a point grid composed of relatively larger points should not have a greater chance of intersecting a reference area than points on a point grid composed of smaller individual points.

An unbiased counting rule for point-area intersections solves this issue. Before starting the point counting process, we select a dimensionless "point-on-the-point," that is, any single arbitrarily chosen point on the point, such as the point where the lower and left arms of the point come together (see arrow on fig. 4.2). Points should be counted if and only if this dimensionless point intersects the reference area. Provided a consistent dimensionless point is used throughout a study, bias arising from the size of the point will be avoided, and only the magnitude of the reference area influences the number of point-object intersections.

6 . . .

VOLUME BY CAVALIERI
POINT COUNTING

- Cavalieri Principle with Point Counting to Estimate V_{ref} from Processing Tissue Sections
- Importance of Using the Correct Section Thickness, t, for V_{ref} Estimation
- Validation of Cavalieri Point Counting Stereology Using Archimedes Principle.

In previous chapters, we introduced the Cavalieri principle as an unbiased and efficient estimator of a 3-D reference volume, V_{ref}, for any region of organic tissue, with no assumptions about shape, size, or orientation. This approach takes the product of the sum of the areas (ΣA) on the cut surfaces and the distance (T) between the cut surfaces to estimate the V_{ref}, according to equation (6.1), $V_{ref} = \Sigma A \cdot T$. We also demonstrated that a point grid with known area per point $[a(p)]$ provides a probability-based probe to estimate the reference area, A_{ref}, on the cut surfaces of 2-D sections. In this chapter, we bring Cavalieri and Delesse principles together for a practical, efficient, and unbiased method to estimate V_{ref} for any anatomically defined reference space.

The schematic in fig. 6.1 illustrates the point counting approach first applied to volume estimation by Gundersen and Jensen (1987). The first plane of the section passes through the reference space at a random location, with subsequent planes parallel and spaced a systematic-uniform distance, T. On each plane, a point grid passes through the reference

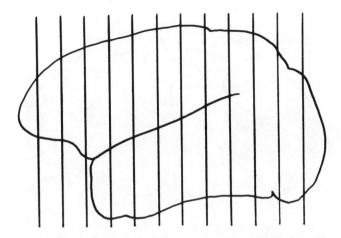

FIG. 6.1. Systematic-Uniform-Random Sections for Cavalieri Volume. Unbiased estimate requires: random first section in the interval T; and all parallel sections.

space, with the sum of points, ΣP, shown intersecting the reference space in the shaded areas. To estimate the actual reference volume, V_{ref}, the area per point, $[a(p)]$, that is, the product of the distances between points in the x and y directions, is adjusted for areal magnification (mag^2). Therefore, the equation for estimation of V_{ref} is shown in equation (6.1).

$$V_{ref} = \Sigma P \cdot [a(p)/\text{mag}^2] \cdot T \qquad (6.1)$$

Equation (6.1) provides an unbiased combination of the Cavalieri principle with point counting stereology to estimate 3-D volume of any anatomically defined reference volume, V_{ref}. This approach is equally effective for estimating V_{ref} from macroscopic tissue, imaging data, or microscopic images, regardless of size, shape, and orientation of the reference volume in question.

Validation of Cavalieri Point Counting Stereology versus Gold Standard

The following example from a published study (Subbiah et al., 1996) validates the Cavalieri point counting method for estimating V_{ref} the

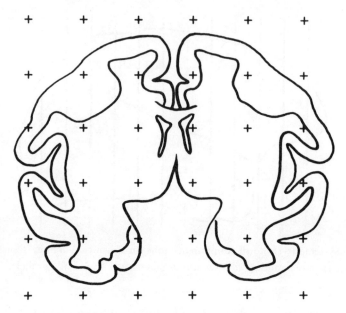

FIG. 6.2. Reference Area Estimation by Point Counting.

gold-standard method for volume determination, Archimedes principle, described in Chapter 1. Single hemispheres (half-brains) from 30
human brains were collected by the autopsy service at the Johns Hopkins University School of Medicine (Baltimore, MD) and fixed in formalin for an equivalent period of time. For measuring the total brain
volume, V_{ref}, each brain sample was submerged in a large tub filled
with water and fitted with an overflow valve and calibrated measuring flask. After each brain sample was placed in water, the amount of
overflow into the flask was recorded. For estimating V_{ref} by Cavalieri
point counting stereology, each formalin-fixed hemisphere was cut in
the transverse (coronal) plane into about 18 parallel, 1.0-cm-thick slabs
using a tissue slicer (fig. 6.2).

Sections were sampled in a systematic-random manner, as follows.
The first cut was made at a random location in the first 1.0-cm interval from the frontal pole, and subsequent cuts were made at 1.0-cm
intervals through the entire sample. Actual data collection was carried out on high-contrast digital images of the anterior face of each
slice. Using a point grid with $a(p) = 180.5$ mm^2, the total number of
intersections between points (+) on the point grid and brain tissue

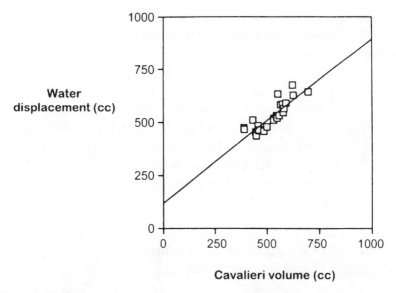

FIG. 6.3. **Archimedes Validation.** Strong correlation between total V for human brain by Archimedes water displacement and Cavalieri point counting methods; $y = 0.774x + 120.4; r = 0.86$.

were counted. Finally, equation (6.1), based on the Cavalieri point counting method, was used to estimate the total reference volume, V_{ref}, of the brain. As shown in fig. 6.3, the same volume measured using Archimedes principle closely agrees with estimates of V_{ref} by the Cavalieri point counting method.

This experiment shows that volume estimation using the Cavalieri point counting method is strongly correlated to volume measurement by a gold-standard method. Because of air bubbles in the hollow ventricles of each brain sample, the correlation line between volumes by each method does not pass directly through the origin; that is, the main source of residual variation in the correlation arises from the water displacement method. In addition to using 2-D slices through an anatomically defined 3-D volume of fixed, frozen, or fresh biological tissue, equation (6.1) is directly applicable to estimate V_{ref} from 2-D slices obtained by in vivo imaging, for example, magnetic resonance imaging (MRI) and computerized axial tomography (CAT). For these studies, the correct slice thickness, T for equation (6.1), is the average linear distance between 2-D image slices.

Analysis of all sections through a volume leads to a calculation of V_{ref} for any anatomically defined reference space. Sampling a known fraction of sections with a random start, for example, every sixth section with a random start on sections 1 to 6, leads to an unbiased estimate of V_{ref}. For each case (individual, animal, subject), a new random starting section is selected for the initial sampling start point, and a new systematic-uniform-random (SUR) sample of sections is selected through the reference space. The approach is mathematically unbiased, because as the level of sampling increases, that is, the number of sections through a reference space increases, the estimate of V_{ref} progressively converges the true value of V_{ref}.

Estimation of V_{ref} Using Cavalieri Principle on Stained Tissue Sections

For the majority of research studies in the biosciences, the estimate of V_{ref} is carried out on microscopic objects on sections following some degree of volume changes required for microscopic observation. These volume changes may arise from the process of dying (agonal changes), perfusion or immersion (drop) fixation in dehydrating agents (alcohol, aldehydes), and alteration in tissue volume as a result of further tissue processing, such as embedding, sectioning, staining, and coverslipping as required for microscopic observation. For estimating V_{ref}, the correct value for T in equation (6.1) is t, the average section thickness after completion of all tissue processing, rather than the microtome setting, also known as the block advance. Typically, use of the block advance in equation (6.3) is a mistake made under the belief that this quantity will estimate the volume of the reference space in the living organism. Rather, this approach calculates an unusual transitional volume, the volume of the reference space on the sectioning block. One must appreciate significant changes in tissue volume caused by a variety of sources, including agonal changes associated with death, shrinkage due to fixation with preservatives (aldehydes, alcohol), embedding in paraffin, heat, dehydration, and so forth, all of which change the tissue volume before tissue sectioning. The only way to quantify the reference volume of tissue in vivo is to base estimates on in vivo images. For brain tissue, the difference between reference volumes in vivo versus after final tissue processing varies in the range of 60% to 70% (Manaye et al., 2007).

When the total number of serial sections through a reference volume is used to quantify V_{ref}, the value T for equation (6.1) equals to the average postprocessing section thickness, t. However, it is never necessary to quantify more than about 8–12 sections through any reference space, regardless of size, shape, and orientation. When the number of sections through a reference volume of interest exceeds 12, a subsample of 8–12 sections is selected by sampling every kth section from the total number of sections that contain the reference area. For example, if the reference volume is present in 40 sections, then every fourth section should be selected in a random start, that is, the first section in the interval of sections 1–4 and every fourth section thereafter, to generate an systematic-random sample of 10 sections. In this case, T is calculated as the product of t, the average postprocessing section thickness, and k, as shown in equation (6.2):

$$T = k \cdot t \tag{6.2}$$

Substituting equation (6.2) into (6.1),

$$V_{ref} = \Sigma P \cdot [a(p)/\text{mag}^2] \cdot k \cdot t$$

Exercise. Tissue was obtained postmortem following perfusion fixation with 4% paraformaldehyde and postfixation in the same solution for 6 hours. Afterward, frozen sections were cut using a knife blade attached to a microtome at an instrument setting (block advance) of 40 µm. The resulting four sections were stained to reveal the anatomical boundaries of the reference volume, which was contained within three sections (shaded region in fig. 6.1). After the slides were stained to reveal the boundaries of the reference space, which required dehydrating with alcohol and heating, the slides were coverslipped using standard mounting media. Measurement of the average postprocessing section thickness t revealed a value of 17.5 µm. A point grid with $a(p) = 625$ mm^2 was placed at random over the three sections containing the reference volume. Point counting carried out on all three sections at a linear magnification of 1500× found 14 inter-

TABLE 6.1. **Parameters for V_{ref} estimation**

t (um)	ΣP	[a(p)] (mm²)	Magnification	k	V_{ref}
17.5	14	625	1500x	1	?

sections between points (+) and the reference space on three sections. Calculate the total reference volume, V_{ref}, using the Cavalieri point counting method and equation (6.1). Table 6.1 summarizes the parameters required to calculate V_{ref}.

$$V_{ref} = \Sigma P \cdot [a(p)/mag^2] \cdot k \cdot t \qquad (6.3)$$

Converting the parameters in table 6.1 to the same units and then substituting into equation (6.3) results in

$$V_{ref} = 14 \cdot [625{,}000{,}000/1500^2] \cdot 1 \cdot 17.5$$
$$= 14 \cdot 278 \cdot 1 \cdot 17.5$$
$$= 68{,}110 \ \mu m^3$$

Correct Choices for Section Thickness

The correct value for t in equation (6.3) is the final section thickness, that is, after the sections are stained and coverslipped on glass slides, as measured by high magnification (oil immersion 60–100x, numerical aperture [NA] 1.3–1.4). This value for t will reflect the condition of the tissue at the time of microscopic observation. Sections cut after freezing or paraffin embedding typically shrink to about 50–60% of the setting of the sectioning instrument (microtome). Note that using the block advance to calculate V_{ref} in the previous exercise would result in a markedly different result—155,556 μm^3—for the volume of the schematic reference space in fig. 6.1. A second reason for using only values after tissue processing is that other estimates on postprocessed tissue assume that the reference volume is estimated following the same amount of shrinkage, for example, estimation of the total number of cells as the product of volume density (cells per unit volume) and the reference volume. Therefore, to avoid introducing errors and to maintain data to compare with other published values, use the final section thickness to estimate reference volume.

7 . . .

ACCURACY AND PRECISION

- • The Difference between Accuracy and Precision
- • Methodological Bias Causes Systematic Error from the True Value
- • Comparison of Area Measurement versus Area Estimation of Accuracy, Precision, and Efficiency

In colloquial speech, the terms *accuracy* and *precision* tend to be used interchangeably; however, in scientific usage, these terms carry different connotations. To appreciate the differences between accuracy and precision in unbiased stereology, we should first consider the definition of *unbiased*: For a given parameter, as sampling increases at each level of the sampling hierarchy, the estimate of the average value of the parameter for a given group converges on the expected (true) mean value for the population. Take, for example, a particular parameter for a specific group such as the total number of brain cells in rats of a given strain, age, and gender. If analyzing more animals, more sections though the brain, or more locations within sections causes the mean estimate for the sample to converge on the true mean value for the population, then the method used to obtain the sample estimate is unbiased. On a theoretical level, use of an unbiased method always leads to *accurate* results; however, those results may be more or less *precise*, depending on the intensity of sampling effort. With high sampling effort, the results for each individual analyzed will be more precise.

Imagine that you wanted to estimate the total number of letters in one shelf of books in a library. One could obtain this estimate by sampling every *10th* book; every *100th* page within every 10th book sampled; every *10th* line on every 100th page sampled; and every *5th* word in every 10th line sampled. Let's say using this scheme you count a total of 150 letters on the bookshelf. An unbiased estimate of the total number of letters on the shelf will be the product of the number of letters counted times the reciprocal of all the sampling fractions at each level of the sampling hierarchy—$150 \cdot 10 \cdot 100 \cdot 10 \cdot 5$—which equals a total of 7.5×10^6 (7.5 million) letters. This estimate may or may not be sufficient for your purposes.

Let's now assume that there are two different bookshelves, and you want to compare the average number of letters per book between them. Using the same unbiased sampling scheme as before, you find that the average number of letters per book on the two bookshelves is statistically different; in this case, the level of precision in your sampling scheme is sufficient to support your conclusion. However, what if the results are not statistically different? Before concluding that the estimates of letters per book on the two bookshelves are indeed the same, you may wish to repeat the analysis with a higher level of precision; that is, decrease the fraction of pages sampled from every 100th to every 50th page. Repeating this analysis may or may not lead to a statistical difference, but the unbiased sampling method will lead to an accurate estimate of the total number of letters, regardless of the level of precision used to sample the objects in question.

For comparison, suppose that you repeat this exercise using a biased method. For example, you assume that the books on each shelf represent identical models of one another. You may then count all the letters in one book (high precision) or count all the letters on a fraction of the total pages (low precision), and multiply these estimates by the total number of books on each shelf. As with the unbiased (model- and assumption-free) method discussed previously, your comparison between the bookshelves may or may not show a different total number of letters; however, in this case, your conclusion will be forever dependent on the validity of the model-based assumption that all of the books are identical.

The use of assumption- and model-free methods leads to an accurate estimate of a sample parameter because the mean value of parameter estimates in the sample will converge on the true or expected value of that parameter for the population. Again, increasing the level of precision may be necessary to obtain statistical support for any conclusions about group differences for treatment and control subjects. With biased methods, the results for the sample parameter will converge at a value that is an unknown (and unknowable) distance from the true value for that parameter at the population level. More or less work (sampling) to estimate a parameter using a biased approach will change the level of precision in the estimate but cannot overcome the bias in the method; once it is present, bias cannot be measured or removed by increasing precision. Thus, unbiased and bias methods will always generate accurate and inaccurate results, respectively, regardless of the level of precision.

With regard to precision, estimates from an unbiased method applied to a single individual, section, or probe location in the x-y plane may be used to make an unbiased estimate; the estimate lacks sufficient precision (reproducibility) to be useful. Increasing precision by sampling more x-y locations on a single section, more sections from a single individual, and more animals from the same group (e.g., control, treatment) will increase the precision of unbiased estimates. Thus, in contrast to accuracy, precision depends on the level of work (sampling) exerted to estimate the parameter in question.

The targets in fig. 7.1 illustrate the central difference between accuracy and precision. The center of each target signifies the central tendency (mean) for a particular parameter, that is, total number of objects (cells) in a defined reference space of interest in tissue. The upper and lower targets on the left show the expected values for an unbiased (accurate) method; on the right, the upper and lower targets show the expected results for a biased (inaccurate) method. A biased method, for example, includes the use of Euclidean geometry to quantify cell volume, that is, volume $= 4/3 \cdot \pi \cdot r^3$, under the false assumption that the cells of interest represent spheres of uniform size. As indicated on the right margin of the image, using either biased or unbiased methods may generate results with high or low precision, depending on whether the data collector uses greater or lesser amounts of time and effort (work), respectively, to sample the tissue. A low amount

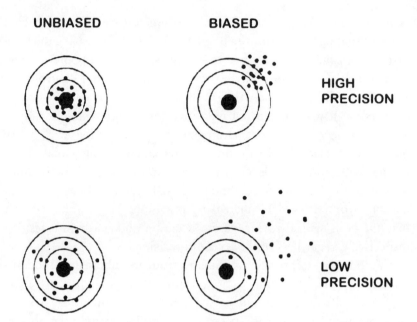

UNBIASED **BIASED**

HIGH PRECISION

LOW PRECISION

FIG. 7.1. Accuracy vs. Precision. Center of targets indicate expected (true) value for the parameter of interest. Precision depends on amount of sampling effort and is independent of accuracy. Thus, unbiased methods give accurate results, regardless of whether high (*upper left*) or low precision (*lower left*).

of sampling effort could analyze only a few cells (<20), whereas high sampling effort would base the same estimate on counting between 100 and 200 cells.

The main idea is that, using either a biased or unbiased method, increasing the work effort by sampling more objects causes the central tendency (mean) of the sample estimate to cluster more tightly *around some value*. Only with unbiased (accurate) methods will that clustering focus on the true value; with a biased method, increasing the work effort will cause the results to converge on a point some unknown (and unknowable) distance away from the true value, otherwise known as systematic error. With biased methods, increasing the work indeed lowers the variance (from lower right to upper right) and causes the data to be more highly clustered; the problem is that the results will more tightly cluster around the wrong point, that is, become more precisely inaccurate. There are no so-called correction factors available to eliminate this systematic error. Because at the start of a study the data collector does not enjoy the benefit of knowing the expected value (otherwise why do the study?), the best precaution against collecting specious data is to use unbiased methods.

Rather than estimate a given parameter to a high level of precision, the goal of organic stereology is to determine whether statistical differences exist to support a given hypothesis, for example, differences between mean values for one or more groups (control vs. treatment). The target and points on the lower left illustrate this optimally efficient strategy—to sample just enough to show group differences, without wasting effort to oversample the tissue to an unnecessarily high level of precision, as illustrated by the upper left target in fig. 7.1. The next section demonstrates quantifying areas using two methods that have potentially equal in accuracy and precision; however, to achieve comparable accuracy and precision, one method must lose efficiency—the amount of precision units captured per unit time of analysis.

Comparing Area Measurement versus Area Estimation

Before applying point counting to estimate area—to measure a reference area, A_{ref}—scientists relied on a machine known as a polar planimeter, a simple hand-held device with a meter that integrates a closed planar area while an operator outlines the periphery of that area. Digital planimetry provides a similar estimate of planar areas by pixel counting within an image digitized into picture elements (pixels). This technique uses the product of two quantities, (1) the area of one pixel, A_{pixel}, and (2) the total number of pixels within the outlined area, N_{pixels}, to calculate a planar area outlined using a manual planimeter. Numerous studies across a wide range of disciplines have compared area measurements using planimetry versus area estimation by point counting stereology using accuracy, precision, and efficiency (Gundersen et al., 1981; Van Vré et al., 2007; Acer et al., 2008).

As shown in fig. 7.2, Gundersen and colleagues compared the accuracy, precision, and efficiency of two methods—pixel counting and point counting—to quantify several areas associated with basal cells in the normal and pathologically altered nasal epithelium. Area esti-

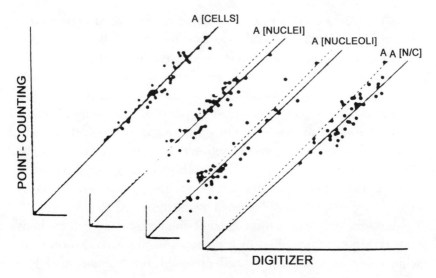

FIG. 7.2. Area Estimate Correlations. Results for area estimations by point counting (*y*-axis) and pixel counting with digitizer (*x*-axis). Both methods equally accurate in theory; however, lower efficiency (more time) required pixel counting to achieve equal accuracy as point counting.

mates using point counting stereology were made on two-dimensional profiles of cells, nuclei, and nucleoli. Regarding accuracy, results between these two morphometric approaches were virtually identical, with the proviso that sufficient time was available for a trained user to carefully outline the reference area for the pixel counting measurement. Although the pixel counting method gave higher levels of precision than the point counting method, this difference in precision for the two methods was negligible in light of the high biological variability from one individual subject (animal, case, subject) to the next. Because the goal is to reduce the variability for the group mean, the higher measuring precision of pixel counting provides little benefit in biological studies where high levels of biological variation exist between individuals within the same group.

Regarding efficiency, the additional time required for accurate measurements using the pixel counting approach dramatically reduced efficiency of that approach, as measured by precision per unit time of measurement. Careful examination of the pixel counting method identified the reason for the lower efficiency of this approach using the point counting technique. For point counting, one uses a combination of retina to visualize the point and brain to determine, P, the number of points that intersect the reference area. In contrast, pixel counting uses the same eye-brain combination to outline the boundary of the reference space but also required a trained hand to carefully outline the boundary. The pixel counting approach depends on this eye-brain-hand dynamic for an accurate measurement of a reference area. An equally accurate estimate of the same reference area by the point counting method does not require the hand dynamic; thus, the point counting method is several-fold higher in efficiency than the pixel counting method. Another group carried out a similar test of accuracy, precision, and efficiency of point counting and pixel counting to quantify a different reference area. That study reproduced these findings of similar accuracy and negligible differences in precision in the group means and reported dramatic gains in efficiency (11-fold) for the point counting method compared with that for pixel counting (Mathieu et al., 1981). Thus, it is difficult to assign an exact value for the gain in efficiency afforded by point counting over pixel counting. The relative efficiencies of these methods will depend on the geometric properties of the reference area in question, with higher gains in efficiency for point counting expected for more complex reference areas.

In a direct test of the "sample, don't measure" concept, we carried out a head-to-head comparison of point counting and pixel counting methods to quantify a relatively complex reference area, A_{ref}, the neocortex of the human brain (fig. 7.3).

Two technicians of equal training and skill level were instructed to quantify the total volume of this three-dimensionally complex reference volume, V_{ref}, using the Cavalieri method, $V_{ref} = \Sigma A_{ref} \cdot T$ (eq. 6.1). From the same set of digitized images, both technicians quantified A_{ref} for each case, either by counting the number of points hitting the neocortical area (estimation by point counting) or by outlining the boundaries of the neocortical area using a digitizer tablet (measurement by pixel counting). Both technicians began the analysis at the same time, under instructions to work at a normal pace and to avoid errors. By the time the technician completed the measurement by

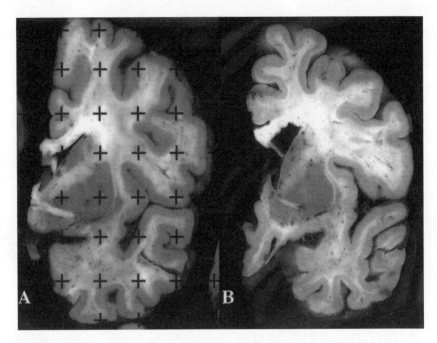

FIG. 7.3. Cortical Volume by Cavalieri Point Counting. Section through postmortem hemispheres of human brains from 60-year-old control (A, with superimposed point grid) and 60-year-old patient with Alzheimer's disease (B). Images from Mouton et al. (1998).

pixel counting on the first image, the technician using point counting had completed the analysis of three cases, with each case consisting of 17–18 images, about a 35-fold increase in efficiency over the pixel counting approach. Once aware of the faster pace of the technician using the point counting approach, the technician using pixel counting increased the pace of data collection on the second image, leading to a clear reduction in accuracy, that is, incorrect inclusion or exclusion of the reference area (neocortex), during the outlining process. By the time the pixel counting technician completed the first case, which took about 4 hours, the point counting technician had easily completed the analysis of 15 cases.

Both technicians then analyzed the set of images using point counting stereology, with *inter*-rater and *intra*-rater reliabilities greater than 99%. The result of this analysis agrees with those reported by other

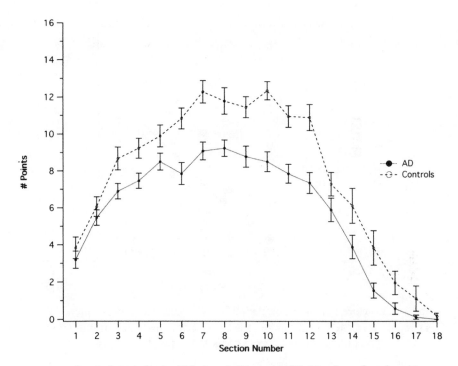

FIG. 7.4. **Cortical Atrophy in Alzheimer's Disease (AD).** Number of probe-object intersections (*y*-axis) for the reference area (neocortex) on 1-cm-thick serial sections shown for the first section in frontal pole to the last section in occipital pole (*x*-axis). Different volumes from integrated areas under curves show 23% loss of neocortex volume for patients with Alzheimer's disease (*lower line*, *n* = 34) compared with that for age- and gender-matched controls (*upper line*, *n* = 19). Data from Mouton et al. (1998).

groups (Gundersen et al., 1981; Mathieu et al., 1981; Van Vré et al., 2007; Acer et al., 2008). In addition to highlighting the practical efficiency of the Cavalieri point counting method, the results from this study (Mouton et al., 1998) found that patients with Alzheimer's disease undergo about a 25% atrophy (volume loss) in the neocortex compared with age- and gender-matched, nondemented controls, as shown in fig. 7.4. Figure 7.5 illustrates results obtained by the point counting method and the Cavalieri principle to estimate total volume of a defined region (human neocortex), and how those data can correlate with results from other modalities such as cognitive performance.

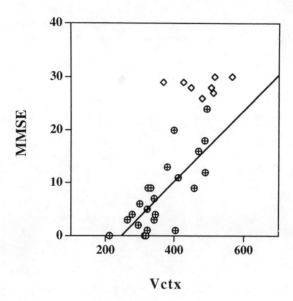

FIG. 7.5. **Structure-Function Correlation.** Loss of total neocortical volume (V_{ctx}) is strongly correlated with cognitive performance on final mini-mental state exam (MMSE) prior to death. MMSE = 0.06 V_{ctx} − 16.28, r^2 = 0.63; $p < 0.0001$).

To summarize, both point counting and pixel counting may be used to quantify 2-D areas on the cut surfaces of 3-D objects, leading to results with comparable accuracy and precision, provided time and care are taken using the pixel counting method, where excessive haste to increase efficiency leads to reduced accuracy. The point counting requirement for only a retina-brain connection, and lack of a "hand dynamic," renders this approach generally more efficient for simple biological structures and remarkably more efficient for relatively complex ones. In chapter 13, we revisit this issue, named "Do More, Less Well" by the great Swiss stereologist Ewald Weibel, with regard to how many animals, how many sections, and how many objects should be analyzed to maximize studies for optimal efficiency.

As for all probes in the field of organic stereology, area estimation by point counting relies on probability theory. The following section discusses another example in which probability theory formed the theoretical basis for estimation of stereological parameters.

FIG. 7.6. Buffon Needle Problem. George LeClerc Buffon proposed the needle problem in 1777.

The Needle Problem

In 1777, George LeClerc Buffon (1707–1788) posed the following question at a meeting of his fellow members of the Royal Academy of Sciences in Paris, France (fig. 7.6):

> What's the chance that a needle dropped on a parquet floor will cross the lines on the floor?

According to the practice of the day, Buffon also provided the solution: the probability of an intersection is directly related to the length of the needle and the distance between the lines on the floor.

A wealthy landowner from an aristocratic family, Buffon was the scientific director of the Royal Gardens in Paris (Jardin du Roi, known today as Jardin des Plantes). As part of his efforts to develop highly organized, decorative hedges, mazes, and other arboreal landscaping, Buffon noted that, despite his most strenuous efforts to create uniformity in the species of plants under his management, those plants retained a certain level of biological variation. Despite his efforts to closely control all environmental factors, such as sunlight, seeds, soil, and irrigation, these biological differences persisted. This observation combined with his botanical research, indicating that new species of plants could be created by cross-pollination, led him to speculate in his scientific work, *Histoire naturelle* (*Natural History*), that laws of nature, rather than divine intervention, drove biological diversity.

In addition to strong interest in biological sciences and philosophy, Buffon possessed an intense interest in mathematics, particularly the emerging field of probability theory. By combining these apparently disparate interests, Buffon endeavored to apply probability theory to test his hypothesis that biological variation fell under the control of natural forces. Like Cavalieri in the seventeenth century, Buffon realized that assumption- and model-free methods were necessary to quantify variation in the surface area of plants that exist as arbitrary-shaped, non-Euclidean objects. The needle problem (fig. 7.7) showed for the first time that the laws of probability could be applied to scientific problems in the late eighteenth century and to a wide range of hypotheses in organic stereology today.

As for estimating the planar area using a point grid, there are only two possible outcomes in Buffon's needle problem: the needle either crosses the lines on the floor ($P = 1$) or something else happens ($P = 0$). Taking one of the quantities as a constant, that is, fixing either the length of the needle or the distance between the lines, and allowing the other quantity to vary, allows one to determine the mean value and variability of the unknown quantity by carrying out this experiment.

Estimating the distance between the lines on the floor using a needle (probe) requires that the intersections between the needle and the line occur in a random manner. If a needle of fixed length L is dropped on a horizontal surface ruled with parallel lines with distance $d > L$ apart, the probability of intersections is related to the value of d, with no further assumptions (fig. 7.7). Randomness of the needle-line intersections ensures that the angle (denoted as theta, Θ)

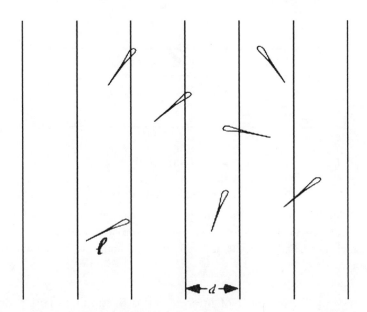

FIG. 7.7. Schematic Needle Problem. Linear probes of length, l, probe lines on a grid separated by a distance, d. The probability of probe-grid intersection, $P(I)$, is directly related to l and inversely related to d [i.e., $P(I) \propto l \cdot (1/d)$].

from the ends of the needle relative to the horizontal lines is randomly distributed [0, 180), as shown in fig. 7.8.

Biological applications of the needle problem to estimate surface area and length assume isotropic probe-object intersection; that is, intersections between the probe and the organic, arbitrary-shaped objects of interest occur with a probability that is equal for all orientations in space. This requirement for isotropic probe-object intersections is met when either the objects of interest in the tissue *or* the geometric probe is isotropic. One example of an isotropic shape is the surface of a sphere. Unfortunately, few if any examples exist of populations of biological objects with an inherently isotropic shape. If such populations did exist, there would be no need for tissue rotation before sectioning; however, essentially no population of organic objects possesses inherent structural isotropy; that is, all populations of biological objects are anisotropic.

The requirement for isotropic probe-object intersection was initially met by random orientation of the tissue that contained the sur-

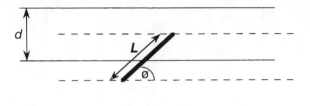

$$p\ (I) = (2/\pi) \cdot (L/d)$$
If p(I) = I/N, then
$$L = (\pi/2) \cdot (I/N) \cdot d$$

FIG. 7.8. **Mathematic Needle Problem.** Number of intersections (I) is used to estimate average length of needle. Random toss of the probe ensures equal probability of angle theta, Ø, between 0° and 180°. In this example, N = total number of tosses.

face area or linear object of interest before sectioning. This process created a series of isotropic-uniform-random (IUR) sections that ensured that all orientations of the object of interest had equal probability of intersection with the geometric probe. After sectioning the previously rotated tissue, probes drawn on transparent acetate sheets were placed over the cut sections, which allowed an operator to count the number of probe-object intersections. For the majority of applications, this requirement for IUR section has been replaced by computer-generated probes. Nevertheless, understanding the need for isotropic intersections requires that we move from the 2-D confines of the needle problem to the 3-D space of organic tissue, as described in the following chapter.

8 . . .

FROM 2-D TO 3-D

- Cutting Tissue Sections Leads to Loss of Dimensions
- Role of Probability Theory in Modern Stereology
- Surface-Area Estimation for Anisotropic Biological Structures

Design-Based Organic Stereology

Organic stereology is the unbiased analysis of arbitrary-shaped 3-D biological objects. Under the influence of wide-ranging evolutionary and ecological forces, biological objects in a defined anatomical space vary in their sizes, shapes, and orientations. Any attempt to model this variation vis-à-vis to Euclidean shapes is likely to fail because biological variation occurs in a stochastic, unpredictable manner, both within a given group of age- and gender-matched individuals, and between groups of individuals treated in different ways, such as untreated versus treated, young versus old, control versus disease, and so forth. By avoiding all assumptions, models, and other sources of biases and uncertainties, unbiased stereology makes accurate estimates of parameters in the small sample of individuals within each group and captures the true variability in these parameters between groups.

The vast majority of organic stereology projects require quantification of magnified images of microscopic objects. Like the anthropologist who changes a culture by the act of studying it, tissue handling and processing to quantify organic objects introduces a number of potential biases and uncertainties into the results. For example, to visualize the specific microstructures of interest, the tissue is typically treated with a variety of powerful reagents and solvents that change the natural structures of the tissues and objects within. Second, to achieve sharp images with sufficient resolution for quantification, a functionally defined 3-D region of interest (reference space) is cut into thin planes (sections) to allow the passage of light, laser, or electron beams through the sections, past various lenses and cameras, and finally to the eye for viewing. Because biological and artifactual factors can potentially skew results from their expected (true) values, organic stereology is design based on necessity: all known sources of variability, bias, and uncertainty are identified, and procedures followed to avoid them.

Sectioning 3-D Objects Leads to Dimensions Loss

The process of eliminating the source of bias begins with the loss of one dimension when cutting 3-D objects into 2-D sections. Sectioning 3-D tissue into 2-D planes leads to partial loss of information about the objects contained within those sections. As fig. 8.1 shows, cutting a 2-D section through 3-D causes a net loss of one dimension (1-D) in the parameters of all microstructures within the tissue. On 2-D sections, 3-D volume (V) appears as 2-D planar areas, 2-D surface area (S) appears as 1-D lines, and 1-D length appears as 0-D points. Potentially more distracting is that this conversion from 3-D to 2-D creates a highly deceptive impression about the number, size, shape, orientation, and connectivity of the objects in the tissue (see fig. 8.1).

S and L of Anisotropic Biological Objects

Biological tissue is composed of arbitrary-shaped organic objects in three possible dimensions. Using a histological section (2-D) to probe the number (0-D) of objects does not fully analyze the 3-D space, because the sum of dimensions in the section (2-D) and parameter (0-D) do not equal 3, the number of dimensions in the tissue. A typi-

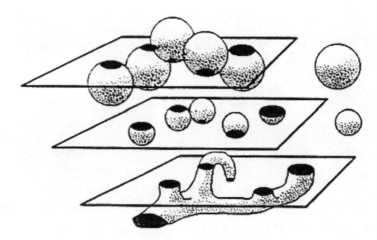

FIG. 8.1. 2-D Section of 3-D Objects. 2-D sections give misleading spatial information about 3-D objects.

cal histological section also does not show information about all orientations of the objects in the 3-D tissue. A natural quality of organic objects is that their shapes lack symmetry; in more mathematical terms, biological objects lack *isotropy*. The classical example of an isotropic shape is a sphere, which has the same appearance regardless of the viewing angle. Because, essentially all organic objects lack isotropic shapes, they are termed an*isotropic*. An example of anisotropic objects is shown in fig. 8.2 in which the objects' orientation on the cut surface of a section is highly dependent on the angle of sectioning.

Two of the four first-order stereological parameters, surface area (*S*) and length (*L*), require special consideration because the inherent anisotropy of biological objects strongly influences the quantification of these parameters on 2-D tissue sections. Anisotropy of biological objects does not affect estimation of the other two first-order stereology parameters, volume (*V*) and number (*N*), because these parameters are not affected by the orientation of the sampling plane. The underlying reasons for the distinction between the effects of anisotropy on *S* and *L*, but not *V* and *N*, are shown in table 5.1 (chapter 5).

There are three possible dimensions that biological objects can

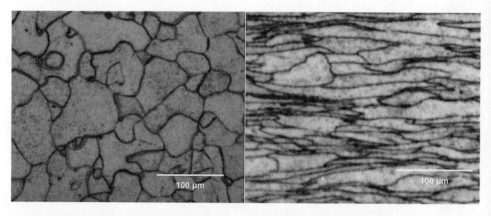

FIG. 8.2. **Anisotropic 3-D Objects in 2-D.** The appearance of nonisotropic (aniso-tropic) objects varies with the angle of observation; bar length = 100 μm.

occupy in tissue. For V and N, the parameter or the probe, respec-tively, covers 3-D of matter. By either quantifying the 3-D volume of an object, or by using a 3-D probe to quantify the number of objects in tissue, the net result is that the 3-D space of the tissue is probed. For S (2-D) and L (1-D), however, neither the parameter nor the probe cover the full 3-D of space within the tissue; therefore, for S and L, possible surface areas or lengths of anisotropic organic objects may reside within the tissue in an orientation that may or may not inter-sect their respective probes. The result is that, before probing S (2-D) and L (1-D) with their respective line (1-D) and planar (2-D) probes, the orientation of either the tissue that contains the biological object of interest or the probe must be randomized in 3-D space.

Because of the inherent anisotropy of organic objects, viewing 2-D sections at different angles exposes different quantities of surface area and length to possible intersections with probes. If steps are not taken to overcome the inherent anisotropy of biological objects when probing for S and L, the results will vary based on the direction of sectioning, even though the sum of dimensions in the probe and pa-rameter equal at least 3. That is, for estimating S using a line probe, the number of intersections per unit length of probe (I_L) will be biased by the plane of sectioning; it will vary depending on whether the tis-sue is cut in coronal, sagittal, horizontal, or other orientation. This

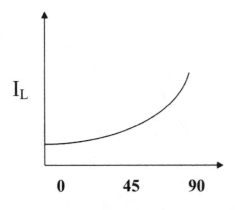

FIG. 8.3. Intersection Probability for Anisotropic Objects. Chance of I_L for anisotropic objects. For the images in fig. 8.6, the probability of a line probe intersection per unit length (I_L) varies with the angle of sectioning from 0° (*left*) to 90° (*right*).

observation is shown by the images in fig. 8.2 in which the angle of sectioning of the same tissue is 90° different (orthogonal) for the images on the left and the right. Figure 8.3 indicates the probability of an intersection per line length, I_L, between the boundaries of anisotropic objects shown in fig. 8.2 and a fixed line probe as the direction of sectioning is rotated through 90°. Since far more surface area of the object is viewed on the image to the right, there is a higher probability of line-object intersections per unit line length of probe (I_L) as the direction of sectioning moves through a 90° angle.

When 3-D tissue is cut into 2-D planes, the appearance of anisotropic biological objects varies in an unpredictable manner (fig. 8.4), as opposed to the predictable appearance of uniformly shaped 3-D Euclidean objects (e.g., spheres, cylinders, cones). Therefore, any method to quantify the S or L of organic objects must be independent of shape assumptions.

Overcoming Anisotropy

To ensure that all surfaces and lengths of anisotropic objects have an equal probability to intersect a line probe or a plane probe, respectively, every intersection between the probe and object must occur with a probability independent of orientation; the probe-object intersections must be isotropic. When this requirement is met, and the geometric properties of the probe are known and fixed—the length of

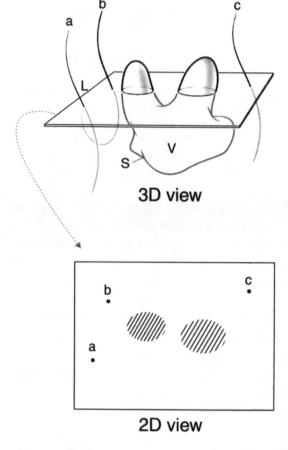

FIG. 8.4. Loss of 3-D Information. Anisotropic 3-D objects in 2-D. One dimension of information is lost by probing 3-D objects with a 2-D section.

the needle in Buffon's needle problem—then the only unknown variable is the magnitude of S and L of the biological objects of interest in the tissue, with no other assumptions about the shape, variation, or distribution. This final point is important because, as mentioned at the start of this chapter, organic structures of biological interest vary in their shapes as a result of evolution, aging, gender, disease, and experimental treatments.

The next section discusses methods to overcome the inherent anisotropy of biological objects for estimation of S and L, which requires that *either* the probe orientation *or* the orientation of the objects is randomized to ensure isotropic probe-object intersections.

Before the current era of computerized stereology, a variety of techniques were developed to randomize the orientation of the tissue prior to estimating the probability of intersections between line and plane probes for surface area and length, respectively. One approach called isotropic-uniform-random (IUR) sampling required complete 3-D randomization of the organic reference space before tissue sectioning. After randomization, tissue sections are cut, and the total number of line-surface intersections is directly proportional to the total surface area, S, and surface density, S_V, of the biological objects of interest. Similarly, on IUR sections, the total number of intersections between a planar surface, the cut surface of the section, is directly proportional to the total length, L, and length density, L_V, of the linear objects of interest. Thus, IUR sampling ensures that all possible orientations of the S or L have equal probability of intersection with their respective probes (see chapter 9 for further details). As for all probes used in organic stereology, this procedure is repeated at about 100 to 200 at systematic-random locations across 8–12 sections through each reference space, as discussed further in chapter 13.

In some cases, a disadvantage of S and L estimation on IUR sections is the difficulty to identify tissue landmarks caused by rotating the tissue 360° before sectioning. This disadvantage can be somewhat reduced using methods called vertical sections (S_V and S; Baddeley et al., 1986) and vertical slices (L_V and L; Gokhale et al., 1990; Stocks et al., 1996). Rather than a grid of straight lines, as used for IUR sections, vertical-uniform-random (VUR) sections use a sine-weighted line called a *cycloid*, as shown in fig. 8.5.

For the VUR approach, the tissue is rotated around a single "vertical" axis through the tissue and sectioned in the horizontal axis, that is, orthogonal to the plane selected as the vertical axis. The term *vertical* refers to the preselected axis, rather than any relationship to gravity—up, down, or horizontal. As for IUR sections, vertical sections are cut with a random start in the reference space, followed by subsequent sections a systematic-uniform distance apart.

VUR sections allow data collection for estimation of S_V and S

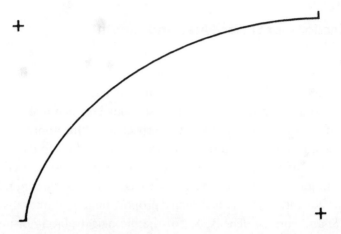

FIG. 8.5. Cycloid Line. A specialized 1-D line-probe used for estimation of 2-D surface area based on Buffon's needle problem. The length of a cycloid line is two times the height measured along the minor (short) axis.

similar to that for IUR sections, though the line probe is slightly more sophisticated for VUR sections. Data are collected using a grid of lines either on transparent overlays (originally) or computer-generated arrays (today). The preferred vertical direction for maximal efficiency is the axis that exposes the maximal surface area or length on the biological objects of interest, which is generally the longest axis of objects as viewed on the cut surface of the section. Because organic objects vary in their orientations in tissue, this selection of vertical axis in no way detracts from the accuracy of the approach but rather requires an approximation that fits the majority of objects of interest (fig. 8.6).

A cycloid is a line with distinctive properties that considers the other two axes of rotation that were not randomized before sectioning. As shown in fig. 8.7, the length of each segment of a cycloid line varies by an amount proportional to the sine of the angle between 0° and 90°.

Figure 8.8 illustrates the correct orientation of the cycloid line in which the minor (smallest) axis is parallel to the vertical axis of rotation. The number of intersections between the cycloid and biological surfaces is directly proportional to the total S and S_v. For estimating the total L and L_v the data collection process is similar, with several important differences. First, rather than counting line-surface intersections after placement of the cycloid grid on the cut surface of the

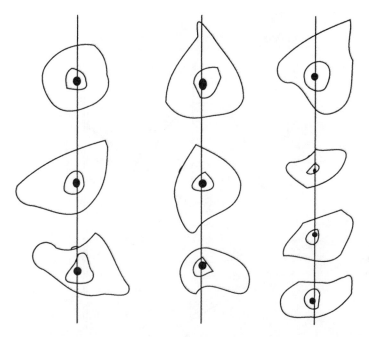

FIG. 8.6. **Vertical Axis of Rotation.** Before sectioning tissue, an arbitrary direction of rotation is chosen for tissue rotation around a vertical axis; for maximal efficiency, this angle should expose the maximum surface area of interest on the cutting plane.

section, for VUR slices, the cycloid grid is (1) placed with the major (longest) axis parallel to the vertical axis of rotation, and (2) the cycloid is focused through the tissue in the z-axis. During this process, the number of cycloid–linear object intersections is directly proportional to the total L and L_v. In combination with VUR sections, the sine-weighted cycloid overcomes the orientation bias associated with the anisotropic surfaces of biological objects in tissue sections (fig. 8.9).

The advantage of this approach relative to IUR sections is the requirement for random rotation around a single axis before sectioning. In many cases, however, this requirement remains a disadvantage because some loss of tissue landmarks occurs after a single-axis rotation; and the requirement for tissue rotation before sectioning precludes using tissue sections, or archival material, that were not rotated around any axis before sectioning. Fortunately, the advent of virtual probes generated by computerized stereology systems allows

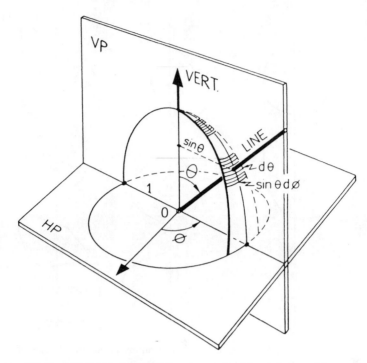

FIG. 8.7. **Vertical Cycloid.** A cycloid line is weighted by the sine of the angle, Ø, of each segment of the line from a selected axis.

for estimating total S and S_v and total L and L_v on tissue sections that may be cut at any convenient axis.

Isotropic Geometric Probes

The requirement for isotropic probe-object intersections requires that either the orientation of the objects in the tissue is randomized before sectioning (e.g., IUR and VUR sections) *or* the orientation of the probe is randomized before intersecting with the object. For the latter case, modern computerized stereology systems allow for software-generated virtual probes that are properly rotated and oriented relative to the objects of interest in the tissue. These virtual probes ensure isotropic probe-object intersections with objects in tissue sectioned at any convenient orientation (coronal, sagittal, or horizontal). Because

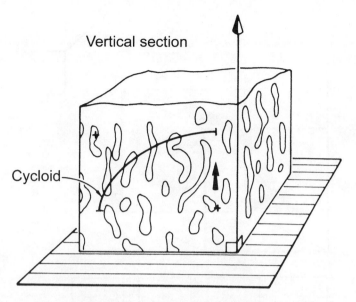

Vertical section

Cycloid

FIG. 8.8. Cycloids and Vertical-Uniform-Random Sections. For surface area estimation, the minor axis of the cycloid is placed parallel to the axis of rotation (vertical) for estimation of surface density (S_V) using the vertical section method of Baddeley et al. (1986) in conjunction with the formula, $S_V = 2 \cdot I_L$, proposed by Smith and Guttman (1953).

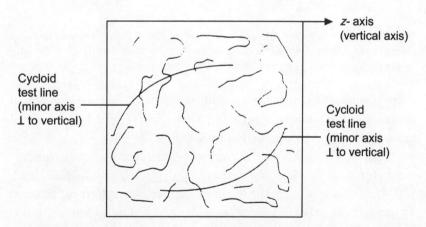

z- axis
(vertical axis)

Cycloid
test line
(minor axis
⊥ to vertical)

Cycloid
test line
(minor axis
⊥ to vertical)

FIG. 8.9. Cycloids and Vertical-Uniform-Random Slices. For length estimation, the major axis of the cycloid is placed parallel to the axis for estimation of length density (L_V) using the vertical slice method of Gokhale (1990) in conjunction with the formula, $L_V = 2 \cdot Q_A$, of Smith and Guttman (1953).

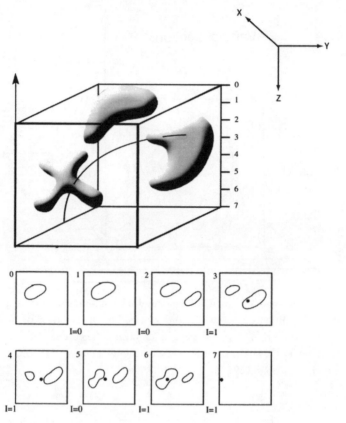

FIG. 8.10. Virtual Cycloids for S_V. The number of probe-object (virtual cycloid) surface intersections provides an unbiased probe for surface density ($S_V = 2 \cdot I_L$), as shown by Gokhale et al. (2004).

there is no requirement for rotating tissue before sectioning, virtual probes allow estimating total S and S_v and total L and L_v on archival tissue not rotated before sectioning.

Figure 8.10 illustrates the computer-generated probes (virtual cycloid) for total S and S_v estimation, as described by Gokhale et al. (2004). In this approach, the sectioning direction is arbitrarily selected as the vertical axis, and software-generated virtual cycloids are oriented with their minor axis parallel to this axis. The cycloid-surface intersections are observed as a dot, which represents a cross section through the cycloid, crossing the surface area of interest, where the number of intersections is directly proportional to the total S and S_v.

For estimating L and L_v, the 2-D planar probe is formed into the surface of a sphere probe (space ball). As an isotropic probe, a sphere contains all integral angles in 3-D space, thus meeting the requirement for all possible directions of probe-object intersections. When this virtual sphere probe is focused through tissue that contains the linear objects of interest, such as fibers and capillaries, the number of sphere-linear object intersections is a direct function of the total number of the linear objects in the tissue (Mouton et al., 2002), as illustrated in fig. 8.10.

Figure 8.11 shows thin focal-plane scanning in the z-axis of a tis-

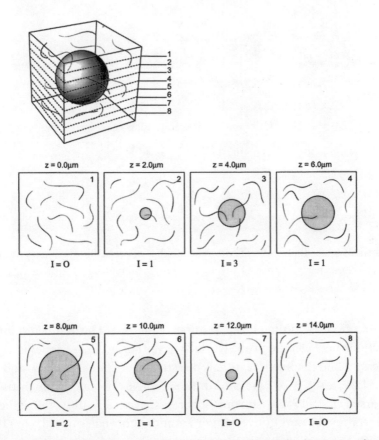

FIG. 8.11. **Space Balls for L_V.** The isotropic surface of virtual spheres provides an unbiased probe for length density ($L_V = 2 \cdot Q_A$), according to Mouton et al. (2002).

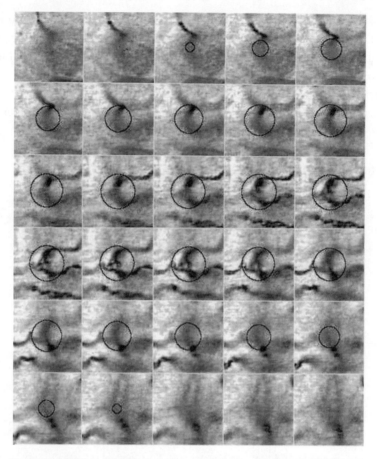

FIG. 8.12. **Space Balls z-Axis Series.** The number of probe-lineal (sphere) object intersections per unit area (Q_A) is directly proportional to the length density $(L_V = 2 \cdot Q_A)$.

sue section containing linear features of interest. The number of inter-sections between the probe (surface of the surface, i.e., circle on cross section) and the linear feature is directly proportional to the L and L_v.

Like a series of planes of different angles passing through tissue (Larsen et al., 1998), isotropic spheres (space balls) allow the tissue to be sectioned in any convenient orientation (coronal, sagittal, etc.) (fig. 8.12). The primary advantage held by spherical probes over plane probes is efficiency. One unique characteristic of a sphere is that the surface contains planes in all integral angles. As a sphere passes through tissue, each plane can potentially intersect with linear features in the tissue with an intersection probability dependent only on the length of the linear features, with no other assumptions (Mouton et al., 2002). To achieve that same level of efficiency as a single tissue pass through, the z-axis with an isotropic sphere probe will require several passes through the tissue, with a plane probe oriented at different random angles during each pass.

9 . . .

SURFACE AREA AND LENGTH

- Equations of Unbiased Estimates of Surface Area (*S*) and Length (*L*)
- Conversion of Densities to Absolute Parameters
- Potential Interpretation Errors due to Scale Dependence

Organic Stereology for Surface Area and Length Estimation

As discussed in the previous chapter, unbiased estimates of surface area (*S*) and length (*L*) require isotropic probe-object intersections. In 1953, Smith and Guttman showed that, in these conditions, the number of probe-object intersection is directly related to total *S* and total *L*, with no further assumptions.

Estimation of Total Surface Area (*S*) and Surface Density (*S*$_v$)

For *S* and S_V, the correct probe must have a minimum of one dimension (1-D) to ensure that the sum of dimensions in the probe (1-D) and parameter (2-D) sums to at least 3. The formula for estimation of S_v is shown in equation (9.1):

$$S_V = 2 \cdot \Sigma I / \Sigma L \qquad (9.1)$$

where

S_v = surface density (1/μm, 1/mm, 1/cm, . . .)

ΣI = total number of line-surface intersections across all sections

ΣL = total length of line-probe used for all sections

To convert the ratio estimator, S_v, to total S, the same approach is used as shown previously in Chapter 4 (eq. 4.6) to convert volume fraction, V_v, to V_{ref}:

$$\text{Total } S = S_v \cdot V_{ref} \qquad (9.2)$$

where

Total S = total surface area in areal units (μm^2, mm^2, cm^2, . . .)

S_v = surface density in ratio units (μm^{-1}, mm^{-1}, cm^{-1}, . . .)

V_{ref} = reference volume in volume units (μm^3, mm^3, cm^3, . . .)

The value of conversion from any ratio estimator to total parameter, such as from S_v to total S, is that the estimate of the total parameter avoids the potential bias from differential shrinkage of the reference space in the ratio estimator. That is, the reference volume of tissue from control and treated groups may not shrink by equivalent amounts during processing; instead, certain groups tend to shrink more or less than others because of tissue responses caused by the treatment, such as aging, lesion, and disease. When this occurs, changes in the volume of the reference space, the denominator in S_V, a ratio estimator, may change in the absence of any real changes in the biological surface of interest. Bias arising from the assumption that all tissue, regardless of treatment, age, presence of disease, and so forth, shrinks to the same extent is called the *reference trap*. The conversion of S_v to total S avoids the potential misinterpretation arising from the reference trap by canceling the bias that arises from nondifferential shrinkage of the reference space. A second advantage to the conversion from a ratio estimator (S_v) to absolute parameter (total S) is that the units for total S (μm^2, mm^2, cm^2, . . .) provide a clearer, more straightforward description of planar area than the units for S_v (μm^{-1}, mm^{-1}, cm^{-1}, . . .).

Estimation of Total Length (*L*) and Length Density (*L_v*)

Similar to estimating S and S_v, the equation of Smith and Guttman (1953) provides an unbiased method to estimate total L and L_v based on the number of isotropic probe-object intersections through a defined reference space. Because the parameter L is 1-D, the minimum number of dimensions in an unbiased probe for L must be 2-D (planar). Therefore, weighting the number of plane-linear object intersections (ΣQ) by the total area (ΣA) of planar probe is used to estimate the L and L_v as shown in equation (9.3):

$$L_v = 2 \cdot \Sigma Q / \Sigma A \qquad (9.3)$$

As described in equation (9.3) for converting S_v to total S, converting the ratio estimator L_v to total L avoids potential bias from the reference trap. The product of L_v and V_{ref} generates an unbiased estimate of total L:

$$\text{Total } L = L_v \cdot V_{ref} \qquad (9.4)$$

where

Total L = total length in linear units (μm, mm, cm, . . .)

L_v = length density in ratio units (μm^{-2}, mm^{-2}, cm^{-2}, . . .)

V_{ref} = reference volume in volume units (μm^3, mm^3, cm^3, . . .)

Expressing total L in length units (μm, mm, cm, . . .) provides an intuitively more straightforward description of length than units of L_v (μm^{-2}, mm^{-2}, cm^{-2}, . . .).

Scale Dependence of S and L

By the mid-1960s, biologists had gained sufficient familiarity with the new methods of stereology to generate data they assumed was based on a stronger mathematical foundation than previous results from methods based on Euclidean assumptions, models, and correction factors. Rather than trying to force populations of arbitrary-shaped organic objects into Euclidean formulas, biologists began to quantify S and L using the equations based on probability theory and stochastic geometry. Though many investigators felt that unbiased stereological approaches were more logically compelling than assumption- and model-based Euclidean methods, reliable data, rather than theoretical discussions, were critically needed to displace the well-ingrained methods of "biased" stereology used at that time.

To provide proof of concept, groups of scientists decided to examine the inter-rater reliability of these approaches using biological tissue. Two particular laboratories, one led by Loud (1968) in the United States and the other by Weibel et al. (1969) in Switzerland, carried out identical studies to quantify the same parameter, the total surface area of endoplasmic reticulum, in the same tissue (rat liver cells) using electron microscopic photomicrographs at different magnifications. However, rather than generating data set in close agreement, the idea that biological parameters of organic objects could be quantified using assumption- and model-free approaches (unbiased stereology) ran into a bit of a glitch.

The inter-rater reliability studies by Loud and Weibel to estimate the same parameter in the same reference spaces did not agree. The results from two stereology laboratories were indeed not even close (table 9.1): Weibel's estimate exceeded Loud's estimate by almost twice (11 m²cm³ vs. 6 m²cm³). After careful comparison of the two studies, the stereologists identified magnification—12,000× for Loud and 80,000× for Weibel—as the only difference between the methods in the two studies.

With the goal of communicating these findings and possibly stimulating a solution from the worldwide community of stereologists, Weibel published these somewhat discrepant results (Paumgartner et al., 1981). For those who believed the value of the new stereology approaches had been oversold and exaggerated by its proponents, these data provided new ammunition. However, Weibel's publication of highly discrepant findings of these studies included a tantalizing clue—the estimates of S_v and L_v at different magnifications showed a direct correlation between resolution and the parameter estimates (fig. 9.1).

For the next half-decade, other groups confirmed these findings as the international community of stereologists discussed and debated

TABLE 9.1. Inter-rater reliability for S_v and L_v.

Magnification	Value	Reference
12,000x	6 m²/cm³	Loud (1968)
80,000x	11 m²/cm³	Weibel et al. (1969)

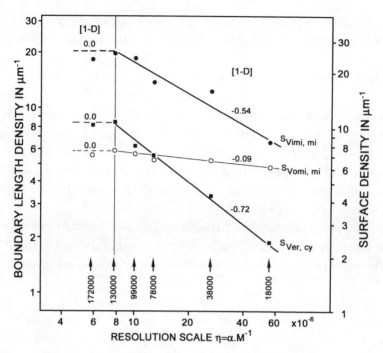

FIG. 9.1. Scale Dependence for S_v and L_v. Results from a comprehensive study by Paumgartner et al. (1981) showing higher stereological estimates of surface and length densities (y-axes) at higher resolutions (x-axis).

their significance. Finally, by staying with this problem long enough, the solution emerged in 1977 at a meeting organized by the Royal Academy of Sciences in Paris, held in the same room and at the same lectern as 200 years earlier, to commemorate Buffon's presentation of the needle problem. A distinguished scientist, the Polish mathematician Benoit Mandelbrot, was invited to speak at this event, and like Buffon two centuries earlier, Mandelbrot posed a question and then provided the answer.

The Coastline of Great Britain

To illustrate this relationship between scale of measurement and quantity measured, Mandelbrot reiterated the coastline paradox first proposed in the middle of the twentieth century by Lewis F. Richardson: What is the length of the British coastline? A person flying in outer space over Great Britain might make one measurement of the surface area and boundary length of this landmass. However, this measurement would be less than if made by a person flying over the country in an airplane, which would be less than a measurement taken while biking around the perimeter, which would be less than a measurement using a caliper applied to each individual pebble around the coastline.

In this example, known as the coastline paradox, the boundary of Great Britain is measured using units 200 km long (fig. 9.2, *far left*), the length of the coastline is approximately 2400 km. With 50 km units (fig. 9.2, *far right*), the total length of the coastline is approximately 3400 km (1000 km longer).

Mandelbrot argued that in fact all of the length and surface-area estimates made of the coastline and surface area of Great Britain were correct because length and surface area are fractal dimensions. Using the same logic, he made the case that the estimates of the total *S* and total *L* of endoplasmic reticulum in the rat liver by Weibel and Loud

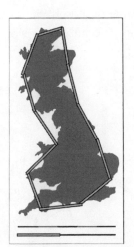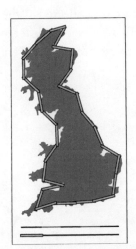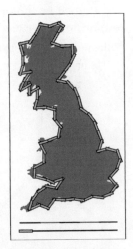

FIG. 9.2. Coastline Paradox. The British coastline is measured using rulers that are 200 km (*far left*), 100 km (*middle*), and a 50 km (*far right*) in length. The more closely a length or surface area is measured, the higher the measured value. Also known as the Richardson effect, after the English mathematician Lewis F. Richardson (1881–1953), who made the initial observation.

were both correct as well. Mandelbrot pointed out that the two studies failed to obtain similar results for data collected at different magnifications, not because of flaws in the stereology methods but because, at higher magnifications, there simply was more surface area and length of endoplasmic reticulum to measure. Both results were correct because S and L are fractal dimensions.

Scale Dependence of S and L

According to Mandelbrot's theory of fractal sets (Mandelbrot, 1977), at a higher resolution more structure is present, sampled, and counted (fig. 9.3). The implications of these observations are important for stereological comparison between studies carried out at different resolutions. For studies carried out at the same resolution to generate mean group estimates (e.g., controls and treatment), the implications of S and L as fractal dimensions are nil. However, by comparing data across different studies carried out at highly disparate resolutions, the effects of Mandelbrot's observations are dramatic: comparison of S and L must be limited to data collected at the same or closely similar resolution; otherwise, the results may differ simply because each was collected at different scales of resolution. This paradox applies only

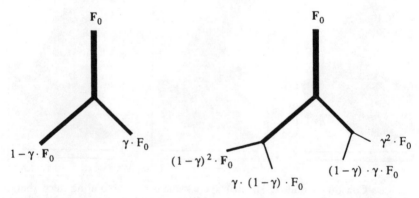

FIG. 9.3. **Fractals.** The figures on the left and right reflect values at low and high magnifications, respectively. Benoit Mandelbrot postulated that at higher magnifications more surface and length are present since surface area and length are fractal dimensions.

to S and L, which are fractal dimensions, and not to the other two first-order parameters, N and V, which are not fractal dimensions.

Needle Problem Exercise

For the IUR section, with lines superimposed over the image, estimate the surface density, S_v, and total surface area, total S, of the objects in fig. 9.4. This image is sampled at random from a total reference volume, V_{ref}, equal to $3.4 \cdot 10^3$ cm^3.

Magnification, $M = 500\times$

Length of one test line = length on the micrograph/M: 5 cm/500 = 1.0×10^{-2} cm = 0.01 cm

Magnification M = 500 X

FIG. 9.4. Surface Area Exercise. The number of line-surface intersections is proportional to the surface density and total surface area. The tissue containing the surface of interest was rotated isotropically prior to sectioning, then sections were cut in a systematic-random manner.

Number of test lines = 3

Total length of test lines = ΣL = 3 × 0.01 cm = 0.03 cm

Total number of intersections with the boundaries = ΣI = 8 + 13 + 8 = 29

$I_L = \Sigma I/\Sigma L = 29/0.03$ cm^{-1}

$I_L = 966$ cm^{-1}

$S_V = 2 \times I_L = 2 \times 966 = 1932$ cm^{-1}

Finally, total S is the product of S_V and the total reference volume, 3.4 × 10^3 cm^3:

Total $S = S_V \times V_{ref} = 1932$ cm^{-1} × 3.4 × 10^3 cm^3 = 6.6 × 10^6 cm^3
= 6.6 × 10^3 m^3

Because the estimate of S_V is based on probe-object intersections at a single, randomly sampled location, further sampling to a total of about 100 and 200 locations is required for surface area estimates with sufficient precision for scientific studies. Regardless of the precision associated with the estimate, because the data were collected with an unbiased probe and random sampling, the estimate of 6.6 × 10^3 m^3 is accurate without further assumptions or qualifications.

10 . . .

TOTAL OBJECT NUMBER

- Profiles in 2-D versus Objects in 3-D
- The Corpuscle Problem and Correction Factors
- D. C. Sterio's Disector Principle

Previous chapters have discussed geometric probability, point grids, and line probes to estimate the probability of an intersection between the probe and the arbitrary-shaped organic objects within a reference space. In these cases, only two possible outcomes exist: either the probe intersects the biological object or it does not. Under the laws of probability, random placement of probes from 100 to 200 x-y locations is sufficient to correctly determine probability of intersection. Once a stable estimate of the probability of a probe-feature intersection is obtained, this information is combined with the known geometric properties of the probe to estimate the unknown magnitude of the parameter of biological interest. In earlier chapters, these concepts were developed for three (total V, total S, total L) of the four first-order stereology parameters. Here, the same concepts are applied to the final parameter, total N.

Objects per Unit Volume (N_V) versus Profiles per Unit Area (N_A)

Total N refers to the total number of discrete objects, for example, cells, in a defined reference space. Typically, total N is obtained from tissue sections stained using histological protocols to visualize a certain population of biological objects. Here, we can draw a distinction

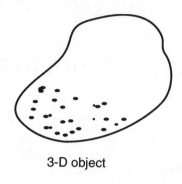

3-D object

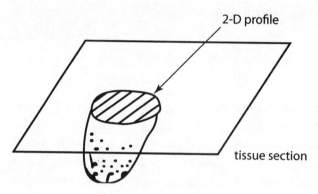

2-D profile

tissue section

FIG. 10.1. **3-D Objects vs. 2-D Profiles.** Profiles are 2-D planes through 3-D objects.

between total number of 2-D profiles per unit area (N_A) on a histological section and the total number of 3-D objects present per unit volume
of tissue (N_V). Published more than eight decades ago by S. D. Wicksell (1925), the Corpuscle Problem illustrates the stereological bias associated with attempting to count 3-D neurons based on the appearance of their 2-D profiles on tissue sections (fig. 10.1).

Figure 10.2 shows a 3-D volume on the left with $n = 6$ arbitrary-
shaped objects; on the right are $n = 10$ profiles that appear on tissue
sections cut through the volume. This schematic clearly reveals that
the number of cell profiles per unit area of tissue, N_A, as viewed on the
2-D tissue section does not equal the true number of cells per unit
volume of tissue, N_V; that is, $N_A \neq N_V$. The next question might be
whether a mathematical formula exists to "correct" this error. That
is, what function might be used to convert the incorrect value (10)

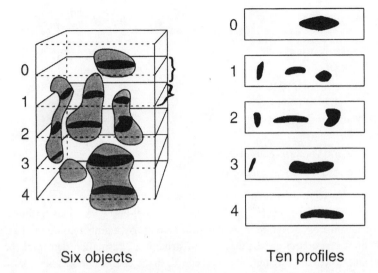

Six objects   Ten profiles

FIG. 10.2. **The Corpuscle Problem.** In 1925, S. D. Wicksell observed that the number of object profiles per unit area on sections through tissue does not necessarily equal the number of objects per unit volume ($N_A = N_V$). The probability that a 3-D object appears on a 2-D section, and therefore counted is a complex function of size, shape, and orientation relative to the plane of sectioning.

into the correct value (6)? Such a function would be useful for converting easily obtained data from histological sections, namely, the number of object profiles per unit area (N_A), into the desired results for the total number of objects in a 3-D volume of tissue (N_V). With his observation of the Corpuscle Problem, Wicksell proposed such a correction factor, the same one discovered independently by material scientists over three decades later (DeHoff and Rhines, 1961).

$$N_V = N_A \cdot D \tag{10.1}$$

where

N_V = volume density

N_A = area density

D = mean particle diameter

This correction formula does successfully convert N_A into N_V, provided that the approach's assumptions are met, namely, that the ob-

Agduhr, 1941

$$N_v = N_A \frac{2m \cdot T - 2r}{Tm \cdot (m + 1)}$$

Floderus, 1944

$$N_v = N_A \frac{T}{T - 2r}$$

Abercrombie, 1946

$$N_v = N_A \frac{T}{T + 2r - 2k}$$

Weibel and Gomez, 1962

$$N_v = \frac{K(N_A)^{3/2}}{\beta(V_v)^{1/2}}$$

Bok and Van Erp Taalman Kip, 1939
and Ebbeson and Tang, 1965

$$N_v = \frac{N_{A1} - N_{A2}}{T_1 - T_2}$$

FIG. 10.3.   Correction Factors. Euclidean-based correction factors use assumption-based coefficients for size (K) or shape (β), or model-based variables (radius). Abbreviations: N_v, actual number per volume unit; N_v, number counted in area A; T, thickness of section; $2r$, diameter of structure; $2k$, smallest detectable part of structure; K, size distribution coefficent; β, shape coefficient; m, number of sections within which the representative cell or nucleus wholly or partly falls.

jects are uniform spheroids of known diameter. For objects with arbitrary and variable shapes present in biological tissues, this correction formula fails because the objects in question do not conform to assumptions and models of the formula. What is the diameter of an arbitrary-shaped object?

In the next five decades, Wicksell's relatively rigorous treatment in the 1920s was followed by attempts by numerous others to introduce variations designed to improve on the correction factor idea (fig. 10.3).

In each of these cases, the correction formula fails for the same reasons it failed when Wicksell first proposed it in the 1920s, which DeHoff rediscovered in the 1960s. For example, the correction factor proposed by Weibel and Gomez (1962) used the volume fraction (V_V) occupied by profiles in a section, to convert N_A into N_V. The problem is that the coefficients K and β depend on arbitrary *guesstimates* of the unknown size and shape distributions of the objects, respectively. In effect, these correction factors simply convert biased estimates of N_A into biased estimates of N_V. Careful consideration of this correction factor idea reveals the underlying flaw in logic—an attempt to *force* non-Euclidean objects into Euclidean formulas. Nevertheless, during

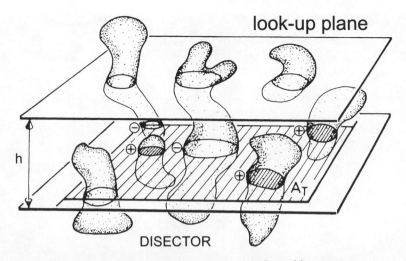

look-up plane

h

A_T

DISECTOR

FIG. 10.4. Discctor Pair. The solution to the Corpuscle Problem is to count "tops" of objects in the same anatomical area of adjacent sections separated by a known height, h. The product of the counting frame area, A_T and h is the volume of one disector (Sterio, 1984).

the first two decades of the modern era of stereology, most studies published about a number of biological objects in tissue used one or another of these assumption- and model-based correction formulas.

The first and, to date, only resolution to the Corpuscle Problem came in 1984 with a principle first published by D. C. Sterio in the *Journal of Microscopy*. This article showed that the true number of objects in a defined 3-D volume of tissue, N_V, can be quantified using a virtual 3-D probe called a *disector*.

The basis of the disector principle is that every discrete object has one topmost point regardless of the object's size, shape, and orientation. The 3-D disector used to estimate total N and N_V consists of an unbiased counting frame of known area, superimposed on one of two adjacent sections, a disector-pair, separated by a known distance smaller than the minimal object height; thus, the disector volume is the product of frame area and disector height (fig. 10.4). The user counts the number of object profiles that appear on the reference section but not on the look-up section. Objects that appear on both top and bottom planes are not counted because their topmost point will not appear between the two sections. In the original demonstration of

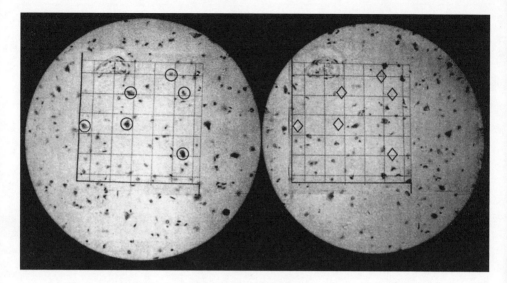

FIG. 10.5. **Physical Disector Pair.** The disector principle was originally proposed for adjacent "physical" sections, as shown here projected next to each other. The circled objects on the left grid "disappear" from the same areas on the right grid (diamonds).

the proof of concept for the disector method, the two physical planes in the disector-pair were tissue sections cut a known distance apart; using blood vessels and other within-section objects, the x-y locations of each section were carefully matched before counting, and both were displayed on the table in front of two separate microscopes projecting down to the surface of a table (as shown in fig. 10.5). When used in conjunction with two thin physical sections, the disector principle is referred to as the *physical disector*.

Unbiased Counting Rules

The disector principle also relies on unbiased counting rules that include inclusion and exclusion planes on the three sides of the 3-D disector, as shown in fig. 10.6. Gundersen (1977) showed that the traditional counting frame, the one in use for decades to count number of blood cells in both clinical and research laboratories, was indeed biased for number (fig. 10.7).

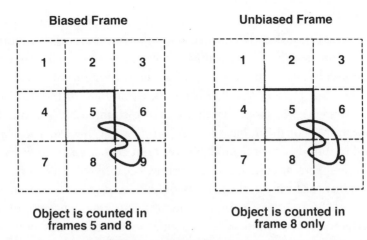

Biased Frame

Unbiased Frame

Object is counted in frames 5 and 8

Object is counted in frame 8 only

FIG. 10.6. **Biased and Unbiased Frames.** Extension of the exclusion lines in the unbiased frame (*right*) avoids the introduction of bias from so-called edge effects.

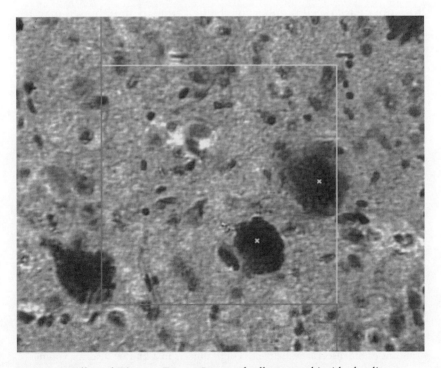

FIG. 10.7. **Cells and Disector Frame.** Image of cells counted inside the disector frame (*middle*) or touching the inclusion line (*far right*). The excluded cell is touching an exclusion line (*far left*).

The frame on the left in fig. 10.7 shows an example of the biased frame in which a slightly concave cell falls within a grid of counting frames. For these frames, the upper and right sides of each frame are inclusion lines, while the lower and left sides of the frame are exclusion lines. Since the cell could hit the inclusion lines of frames 5 and 8, this single cell has a probability of being counted twice using this biased frame. For the unbiased frame on the right, the exclusion lines extend to infinity down from the lower right corner and to infinity up from the upper left corner. For this frame, because the cell hits the exclusion line for frame 5, the single would be counted only in frame 8. The breakthrough here is that no matter how one conceives of the size, shape, and orientation of a single object, using this unbiased frame and counting rules from Gundersen (1977), the object will only be counted once. In combination with the disector principle, Gundersen's unbiased frame and counting rules allowed biomedical scientists to estimate total cell number for the first time in an unbiased manner, without bias associated with the size, shape, or orientation of the objects in the tissue (Sterio, 1984). In practice, users simply count objects of interest either inside the disector frame or touching the inclusion lines, as illustrated in fig. 10.8 by a screen capture from a computerized stereology system (*Stereologer*, Stereology Resource Center, Chester, MD).

Alert readers will notice that *disector* and *D. C. Sterio* share the same letters but in different arrangements. With regard to the true iden-

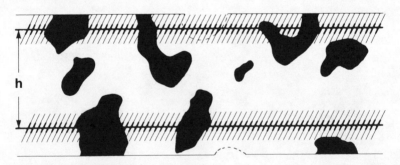

FIG. 10.8. **Optical Disector with Guard Zones.** The optical disector (Gundersen, 1986) uses the disector principle to count objects in a known disector volume within thick sections (typically 30–50 μm before processing), in combination with a guard volume (hatched regions) to avoid sectioning artifacts, e.g., lost caps, caused by the knife blade.

tity of the person responsible for the disector principle, an interesting note at the bottom of the 1984 paper by anagrammatist D. C. Sterio states, "D.C. Sterio is the non de plume of a stereologist who does not wish his/her name to be associated in perpetuity with the method given in this paper."*

Validation

One year after publication of the disector principle by D. C. Sterio, two Canadian scientists (M. Colonnier and C. Beaulieu) used a straight-forward method of gold-standard testing, a known number of phantoms embedded in a transparent matrix, to test the accuracy of all methods to determine object number of tissue sections in use at the time. Included in this validation study were the disector principle, unfolding methods, and studies using a variety of so-called correction factors. Of the various methods evaluated, the only method to produce accurate results was the disector method. As expected for an unbiased method, with increased sampling intensity the estimate of total object number using the disector principle gradually converged on the true number of phantom objects; in contrast, the other methods, including the assumption- and model-based correction method of Colonnier and Beaulieu, misestimated the known number of phantom particles with systematic error (bias) rates up to one-third (32%) of the true value.

Two years after publication of the disector principle by D. C. Sterio, Gundersen (1986) introduced the *optical disector*, an important and novel application of the disector principle. Rather than two physical sections serving as the disector-pair, objects in the disector volume are counted by thin-focal plane optical-plane scanning through the z-axis, through thick tissue sections typically cut at an instrument setting (block advance) of 30–50 μm and processed to a final section thickness of 15–20 μm. Like the physical disector, this approach uses the disector principle to identify objects present on one plane (reference plane) but not on the adjacent plane (look-up plane); however, in

*In deference to the wishes of the author of the disector principle, the true identity of D. C. Sterio will not be revealed here.

the optical disector approach, both the look-up and reference planes are serial optical planes rather than adjacent tissue sections.

The images in fig. 10.8 illustrate the optical disector approach schematically using a series of focal planes through the z-axis viewed orthogonal to the cut surface. The hatched regions indicate the guard volumes at the top and bottom of the section. Because the tissue at the cut surface of the sections may suffer from sectioning artifacts by the knife blade, for example, tops of objects cut off (lost caps) or objects pulled out of the tissue, no cell counting is done in these guard volumes. Using Gundersen's unbiased counting frame and counting rules, the optical disector method identifies the expected number of unique counting objects (e.g., cells, nuclei, nucleoli, middle focal plane), as indicated in fig. 10.9 by a pair of optical z-axis planes viewed through a stained tissue section (West and Gundersen, 1990).

In practice, optical disector counting uses a stack of such focal planes in the z-axis to count the number of unique counting items with a 1:1 stochiometry, that is, 1 counting item = 1 cell. Note the dark nucleoli, a unique counting item for this cell type that appears in the nucleus of the cell while moving from the upper to lower image in fig. 10.10. Optical disector counting with cell nuclei as the counting item through a stack of focal planes in the z-axis is illustrated in fig. 10.11. The observer locates the most in-focus plane for the counting item and only counts objects touching the inclusion planes or inside the disector volume, as shown by white arrowheads in fig. 10.11. All other in-focus counting items, including those touching both the inclusion and exclusion planes, are excluded (fig. 10.11).

Because objects are scanned and counted using thin focal-plane scanning, the optical disector requires counting at high-resolution objectives that allow observation of thin focal planes, ideally 60× and 100× oil immersion objectives with numerical apertures of 1.3 to 1.4 for counting number of cell and subcellular structures, respectively. These high-quality objectives provide the high focal depths and low depths of field necessary for thin focal-plane scanning with the optical disector method through a relatively thick section. Non-oil immersion objectives with lower numerical apertures and magnification less than 60× have focal depths too low for thin focal-plane scanning and

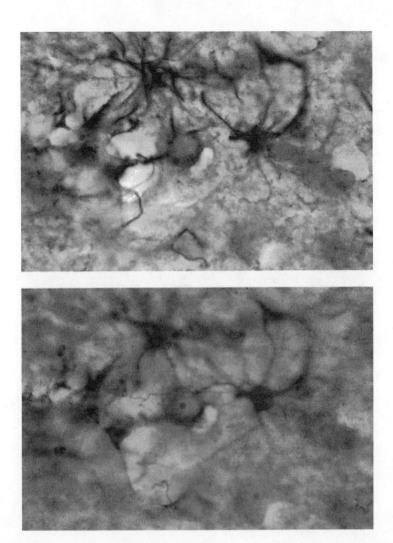

FIG. 10.9. Optical Disector Pair. The optical disector method uses thin focal-plane optical scanning in conjunction with thick tissue sections. Scanning in the z-axis using a thin focal-plane created by high-resolution optics (100× oil immersion, numerical aperture 1.4) causes the cross sections of dark nucleoli to disappear. The calculations for the optical disector are the same as shown in fig. 10.6.

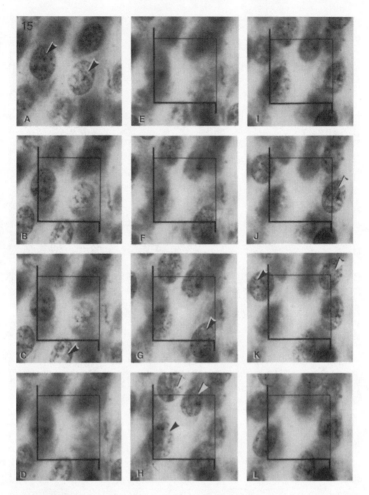

FIG. 10.10. **Thin Focal-Plane Optical Scanning in z-axis.** For the optical disector method (Gundersen, 1986), a stack of optical planes is scanned in the z-axis. When a cell nuclei comes into focus and intersects the disector frame, without touching either of the exclusion lines (black arrow), the cell is counted (white arrow). To avoid sectioning artifacts, no counting is done within the guard zone in top left frame, i.e., the optical planes between the top surface of the section and the top plane of the optical disector.

$$N = \frac{\overset{j}{\Sigma Q^-}}{h \cdot \underset{i\,=\,1}{\overset{j}{\Sigma}} a(frame)} \cdot V(ref)$$

FIG. 10.11. Director Equation for Total N. Objects are counted in an unbiased manner using disectors spaced throughout the reference space. The formula for calculating the numerical density (N_V; number, N, per unit volume, V) is the sum of disappearing cross sections, ΣQ^-, from the German cross section, *Querschnitt*, divided by the total volume of disectors (product of the area of one disector and the total number of disectors). Shrinkage artifacts in the estimate of total N are eliminated in the product of N_V and total reference volume, V_{ref}.

precise determination of section thickness required for use of the optical disector.

To make a stable estimate of the probability of disector-object intersection, disector counting is repeated at between 100 and 200 locations in the *x-y* plane across about 8–12 systematic-uniform-random (SUR) sections through a reference space. Both physical and optical disector methods lead to estimates of neuronal density, N_V. To avoid the reference trap, and thus avoid bias in the event of differential shrinkage of the reference volume, N_V is converted to an absolute parameter, total neuron number, using the following equation (fig. 10.12). where

N = total number estimate for objects in the reference space

ΣQ^- = sum of disappearing cross sections,* i.e., the sum of objects counted within volume of all disectors.

Σa (frame) = sum of the areas for all disector frame (mm^2)

h = height of the disector (mm)

V_{ref} = volume of the reference space (mm^3)

*The sum of "disappearing" cross sections is indicated in the stereology literature as ΣQ^-, from the German, *Querschnitt*, for cross section.

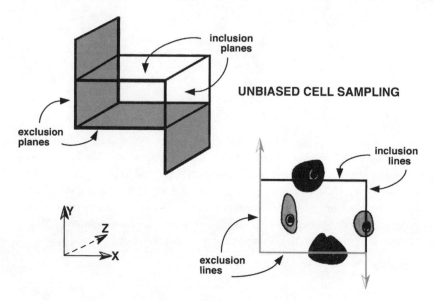

UNBIASED CELL SAMPLING

FIG. 10.12. Exclusion and Inclusion Planes. Number of objects (cells) is counted using counting rules from Gundersen (1977), including exclusion and inclusion lines that become planes when viewed as part of a virtual 3-D disector.

Consideration for $N = N_v \cdot V_{ref}$

With respect to avoidance of the reference trap, any amount of shrinkage that occurred in the denominator of the ratio N_v will cancel in the product of N_v and V_{ref}.* Thus, this formula resembles those previously discussed (e.g., see eq. 4.7) to avoid the reference trap by conversion of a ratio estimator (N_v) to an absolute parameter (total N). Cancellation of bias due to differential tissue shrinkage assumes equivalent methods of tissue processing are carried out for both the N_v and V_{ref} estimates. This requirement is met when, for example, V_{ref} and N_v are estimated on the same tissue sections. However, this requirement is not met if, for example, V_{ref} is estimated on unprocessed tissue (e.g., on macroscopic gross tissue) and N_v is estimated on sectioned and stained slides. Exposure to differential solvents and dehydrating agents, for example, alcohol, during tissue processing for estimation

---

*Similar to the cancellation of hours in the following equation: Number of miles traveled = 55 miles per hour · 10 hours = 550 miles.

of N_V could cause variable tissue shrinkage in the estimates of V_{ref} and N_V; in this case, the tissue shrinkage in the two multipliers would not cancel in the product V_{ref} and N_V. Another factor for consideration with this conversion is that need to include all volumes when making estimates of N_V, whether those volumes contain any objects of interest, which may be difficult in some comparisons of reference spaces between control and treated groups. The following section describes a further enhancement of the disector principle, the fractionator.

The Fractionator

The optical fractionator method (Gundersen, 1986) combines the optical disector principle with an efficient sampling scheme that uses known sampling fractions to scale from a local estimate of N_V to an estimate of total N for the entire reference space. As for the optical disector approach described, the same disector counting procedure is carried out in the z-axis of thick sections, and this process is repeated at between 100 and 200 x-y locations to obtain a stable estimate of the probability of intersections between the disector probe and the biological objects of interest. The difference from the method described above for calculating total N from the formula given in fig. 10.12 lies in the exclusion of V_{ref} from the fractionator equation. In contrast to estimating total N as the product of N_V and V_{ref}, the fractionator method calculates total N as the product of ΣQ^- and the reciprocal of all sampling fractions, as shown in equation (10.2).

$$\text{Total } N = \Sigma Q^- \cdot (F_1) \cdot (F_2) \cdot (F_3) \tag{10.2}$$

where

Total N = estimate of total object number

ΣQ^- = number of objects actually counted

F_1 = 1/ssf = 1/number of sections analyzed/total number of sections

F_2 = 1/asf = 1/area of the disector frame/area of the x-y step

F_3 = 1/tsf = 1/thickness sampling fraction = 1/disector height/section thickness

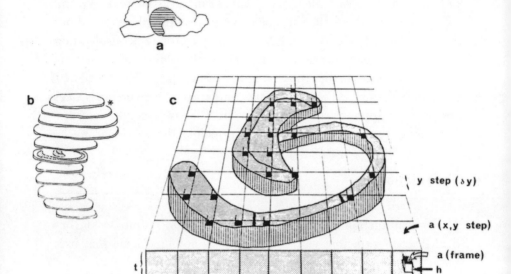

FIG. 10.13. The Fractionator. The optical disector in combination with the fractionator sampling scheme (West et al., 1991). The optical disector method is used to make an unbiased count of objects within a known fraction of the total sections (section sampling fraction), a known fraction of the total area of each section (area sampling fraction), and a known fraction of the section thickness, thickness sampling fraction.

Comparison of Fractionator and $N_V \cdot V_{ref}$ Approaches for Estimation of total N

Gundersen's fractionator method (fig. 10.13) enjoys several advantages over D. C. Sterio's proposed method to find total N as the product of N_V and V_{ref}. First, the fractionator calculation does not involve volume for estimates of either N_V or V_{ref}; therefore, differential changes in the volume of the reference space for different individuals, for example, young versus old, do not affect the final estimate of total N. Second, data collectors need only ask a single question at each x-y location: Is a biological object (cell) of interest present in the disector? If no, this location is skipped and the same question is repeated at the next x-y location; this in contrast with the $N_V \cdot V_{ref}$ approach in which all volume in the reference space must be included regardless of whether the volume contains objects of interest. The practical aspect of this difference is that for the fractionator approach we may identify the reference space based on the presence of the biological objects of interest; for the $N_V \cdot V_{ref}$ method, we must be able to identify the anatomical boundaries of the reference space, regardless of whether biological objects of interest are present. As a result of these advantages over the $N_V \cdot V_{ref}$ method, optical fractionator method has become the state-of-the-art approach for making reliable estimates of total N for biological objects (West et al., 1991; for recent examples of the optical fractionator using immunocytochemical stains, see O'Neil et al., 2007; Manaye et al., 2010; Mouton et al., 2010).

The following chapter provides a further modification of the fractionator approach specifically designed for so-called rare event situations in which the total number of biological is unusually low.

Brain Aging and the Reference Trap

After the publication of the disector principle and the validation of the approach in the mid-1980s, scientists began to use this powerful approach to address long-standing controversies and contentious hypotheses that involved changes in total N. One hotly debated issue at that time surrounded the question of neuron loss in the brain during normal aging. Data based on a wide range of studies for the previous two decades that used Euclidean-based methods (stereometry) were in general agreement. These studies by several leading laboratories in the field of brain aging supported the view that after about age 50 the human brain underwent a clear reduction in neuron density in various areas of the cerebral cortex; furthermore, these changes appeared to coincide with behavioral studies of decrements in various indices of cognitive, sensory, and motor function with onset around about the same age (Brody 1955; Roberts and Goldberg 1976). Since these assumption- and model-based data appeared to provide a neurobiological basis for the age-related loss in brain function, these stereometry-based data were widely reported and defended, despite studies to the contrary

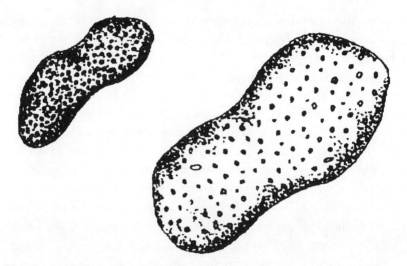

FIG. 10.14. The Reference Trap. Ratio estimators such as density (N_V number per unit volume) can lead to false conclusions if there is differential shrinkage of the reference volume due to aging, disease, lesion, and fixation times.

reporting no age-related changes in neuron numbers (Berlin and Wallace 1976).

Dogma surrounding age-related loss of cortical brain cells continued to unravel with reports by Dam (1979) and Haug (1984) of an inverse relationship between age and tissue shrinkage (volume loss). These studies reported that autopsied brain tissue from older individuals undergoes less shrinkage than that for younger individuals after equivalent periods of aldehyde fixation (fig. 10.14). Because greater shrinkage of the reference space would lead to higher neuron density for younger individuals, this study raised the possibility that the age-related changes in a ratio estimator and number of cells per unit area (N_A), reported by Brody (1955) and others, may simply reflect the age differential in tissue shrinkage, without any changes in the total N. The higher degree of tissue shrinkage in these brains was compared with brains from older individuals. Using unbiased stereology approaches to quantify total neuron number, as opposed to neuron density, several studies by different groups reported that neuron number in most brain regions is stable in the same brain regions reported previously to lose neurons starting around age 50 (Mouton et al., 1994;

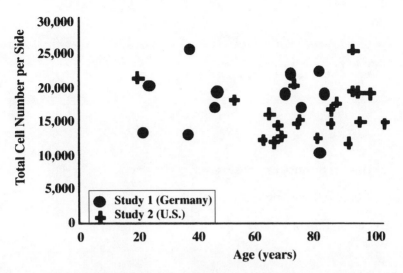

FIG. 10.15. No Change with Age. Contrary to results from numerous studies of age-related cell loss using biased stereology methods, two independent studies carried out in separate countries by different investigators (Mouton et al., 1994; Ohm et al., 1997) obtained similar results showing no age-related cell loss using the unbiased optical disector method.

West et al., 1994; Morrison and Hof, 1997; Ohm et al., 1997; Pakkenberg and Gundersen, 1997; Long et al., 1999; Pakkenberg et al., 2003).

The results shown in fig. 10.15 are from two independent studies of the total N in nucleus locus coeruleus in the human brain (Mouton et al., 1994; Ohm et al., 1997). Together with results from similar studies in humans and experimental animals, these estimates of total N using design-based stereology support the view that age-related neuron loss is likely an exception, rather than the rule, in the normally aging brain (West et al., 1994; Morrison and Hof, 1997; Long et al., 1999). The prevailing view among neuroscientists in the field of brain aging now focus on loss of connectivity or reduced neuronal function, instead of neuron loss, as the most likely basis for age-related loss of brain function.

11 . . .

RARE EVENTS

• The Occurrence of Rare Events

• A Single Sampling Fraction

• Calculation of Total *N*

Estimation of Total *N* for Rare Events

The previous chapter detailed two techniques, the disector principle and fractionator sampling, as unbiased approaches to estimate total *N* of biological objects in organic tissue. For most applications in the sciences, typical applications of these approaches involve counting objects at about 100 to 200 locations through about 8–12 sections in the reference space. For typical studies of this type using computerized stereology systems (chapter 15), the time required to estimate total *N* for one individual is about 1 hour. However, situations may arise in which the expected total number of biological objects in a defined reference is unusually low, so-called rare events, which may reduce efficiency for the object-counting procedure. In response to these situations, a particular stereology design called the rare event protocol (REP) provides an efficient alternative to the standard approach, without a loss of accuracy or precision.

In the case of rare events, with typical sampling approaches, the majority of disector probes will not intersect objects of biological interest. For example, if each of the 8–12 sections contains 20 or fewer objects distributed randomly through the *x-y* plane, then closer spacing between adjacent disector locations is required to increase the probability of disector-object intersections. However, this approach will dramatically increase the time required to analyze each section. The net effect of the high-stringency sampling required for rare events is to dramatically increase the time to estimate the parameters of interest in each subject. For a study with a total of $n = 40$ individuals ($n = 10$ per four group), this enhanced effort makes the process highly time consuming, perhaps threefold higher than for a nonrare event. In this scenario, the REP is an efficient alternative that restores the efficiency for estimating total N in one individual to about 1 hour. The REP, a variant of the fractionator principle (Gundersen, 1986), achieves this gain in efficiency by substituting manual scanning in the x-y-z planes for the more time-consuming stepping with a motorized stage in a systematic-random manner.

Systematic-Random Sampling of a Known Fraction of Sections

In the case for nonrare events, the first step in the REP is a systematic-random sample to obtain about 8–12 sections through a defined reference space. For instance, from a total of 60 sections, selecting every 6th section will generate a desired sample of 10 sections. The first section is sampled with a random start from the 1st to the 6th section (e.g., 5th section) and then every 6th section collected thereafter for a total of 10 sections (i.e., sections 5, 11, 17, 23, 29, 35, 41, 47, 53, 60). This procedure will generate a systematic-random sample of 10 sections with a section sampling fraction (ssf) of one-sixth.

At this point, the sampling and data collection process to estimate total N for rare events diverges from that for nonrare events. Rather than using a stage motor to step in a systematic-random manner between locations in the x-y plane, the full thickness of each sampled section becomes the disector volume, which represents a known fraction, for example, one-sixth, of the total volume. At this point, cells of interest are counting in the x-y-z dimensions of each of the 10 sections by scanning the stage manually through the sampled sections.

Avoiding the Corpuscle Problem

This pattern of manual scanning in the x-y plane is done at low power, typically with a 40× objective, to locate stained objects of interest in the section. Bias from the Corpuscle Problem (Wicksell, 1925) is avoided by, once an object is identified, switching to high magnification to count only the topmost point of objects in the volume of the section. If the object is cut on the first plane of the section, then the topmost point of the object is not present in the section; in that case, the object is not counted. If the object is brought into focus by moving into the volume of the section, the user counts (clicks on) the object, and continues manual scanning at low power through the section. This process is continued until the total volume on all the sampled sections have been scanned at low power, and all cells with their topmost point within the sections counted.

Calculation of Total N

As given in equation (10.2) for the fractionator method, the formula is repeated below in equation (11.1) for calculating total N of rare events.

$$\text{Total } N = \Sigma Q^- \cdot F_1 \cdot F_2 \cdot F_3 \tag{11.1}$$

where

ΣQ^- = total number of objects with topmost point within the volume of sampled sections

F_1 = 1/section sampling fraction (ssf)

F_2 = 1/area sampling fraction (asf)

F_3 = 1/thickness sampling fraction (tsf)

The differences for the REP are the following. First, because the full area on each sampled sections, the area sampling fraction (asf; area of the sampling frame/distance of the x-y step) is equal to 1; and second, because the user counts all objects of interest present in the entire section thickness, the thickness sampling fraction (tsf; height of the disector/section thickness) is equal to 1. Because all objects are counted within a known fraction of the reference space, the total number of objects through the reference space is simply the product of the sum of object tops (ΣQ^-) in the sampled sections and the reciprocal of the ssf, as shown in equation (11.2):

$$\text{Total } N = \Sigma Q^- \cdot F_1 \cdot F_2 \cdot F_3 \qquad (11.2)$$

where

ΣQ^- = total number of objects with topmost point within the sampled sections

F_1 = 1/section sampling fraction (ssf)

F_2 = 1/asf sampling fraction (asf) = 1

F_3 = 1/thickness sampling fraction (tsf) = 1

By eliminating the time-consuming process of sampling hundreds of empty disectors using a stepping motor with a low distance between disectors, the REP dramatically increases the efficiency to estimate total N in rare events. The following section reviews the tissue processing and magnification required by this approach.

Tissue Processing and Magnification Requirements

Section thickness. Generally thicker sections, that is, those cut at an instrument setting (block advance) of 40–50 μm or greater, provide more objects to count per section and, therefore, more efficient estimates of total N with the REP than for a block advance of 30 μm or less.

Staining. Objects of interest must be stained with sufficient contrast, that is, high signal noise, to visualize the objects (signal) against background (noise) at low magnification.

Magnification. The REP requires low- and high-magnification microscope lenses. A low-magnification objective (e.g., 20–40x) is required for users to visualize objects of interest while manually scanning across the x-y planar surface of each sampled tissue section. As for all stereological studies, the ideal magnification is the lowest magnification for users to unambiguously identify the objects of interest. Second, a high magnification, for example, oil-immersion lens (63× or 100×, numerical aperture 1.4) is required for the user to reject objects of interest with a missing cap, that is, topmost point of object cut off at the top of the section. This requirement ensures that only the object with the topmost point present within each sampled section is counted. Provided tissue sections are properly sampled and stained, and the required low- and high-magnification lens are available, the REP increases the efficiency to make unbiased estimates of total object number to approximately the same level as for quantification of non-rare objects on sections using computer-assisted stereology approaches.

To Step or Not to Step

Efficiency is typically expressed as precision units captured per unit time. The REP allows for fractionator-based estimation of total N for relatively rare objects (events) but with remarkably higher efficiency than would be possible with traditional stepping motors. For rare events, the motorized stepping motor reduces efficiency because of the need for exceedingly small steps between disector locations. At what point does the REP increase the overall efficiency of the procedure?

Two major factors that affect the efficiency of the REP are expected total N and relative clustering of those objects in the reference space. The REP approach becomes the more efficient design as the total N in the reference space falls below about 3000 total objects; the farther below 3000 total objects, the greater the gain in efficiency of the REP compared with sampling with motorized step motors. With respect to clustering, the biological objects may be homogeneously distributed through the reference space (low clustering) or distributed in a more heterogeneous distribution (high clustering). The added efficiency of the REP increases as the level of clustering within the reference space increases. The value of this approach in terms of capturing more units of precision in less time, relative to that using motorized step stages to sample tissue, increases as the expected total N falls below 3000 objects and the level of clustering within the reference space increases.

Example

A particular reference space appears on 120 sections that have been cut at an instrument setting of 50 μm. Every 12th section with a random start is sampled for 10 sections (ssf = 1/12). Staining with bromo-deoxy-uridine (BrDU) reveals rare numbers of dark nuclei against a relatively light background. Low-magnification scanning at 40 μm identified 52 cells (ΣQ^-) with their topmost points within the 10 sampled sections. On the basis of this information, use equation (11.2) to calculate the total N of BrDU cells in the reference space.

Solution. Using equation (11.2), the total N of high $S{:}N$ stained objects in the reference volume is 624 cells (total $N = \Sigma Q^- \cdot F_1 = 52 \cdot 1/1/12 = 624$).

Thus, by low-magnification manual scanning through the x-y-z plane of a known fraction of the total volume of the reference space, the REP achieves increased efficiency, with equal accuracy and precision compared with estimating total N using traditional (stepping motor-based) computerized stereology. The efficiency of the REP relative to traditional computerized stereology increases as the objects of interest cluster together within the reference space and decreases as the expected total number of objects of interest approaches about 3000. The improvement in efficiency over traditional stepping motor-based sampling approaches will depend on the number of actual objects and their clustering in the reference volume.

12 . . .

LOCAL SIZE ESTIMATORS

- • Measurement versus Estimation of Object Size
- • Size Estimators Based on Pappus-Guldinus Theorem
- • Number- and Volume-Weighted Size Estimators

Previous chapters discussed methods to estimate the total volume of a region, total V, using the Cavalieri point counting method and the total object volume, total V_{obj}, using point grids for area estimation. This chapter focuses on estimators of local volume, including the volume-weighted mean object volume, MOV_V, using point sampled intercepts (PSI) sampling and the number-weighted mean object volume, MOV_N, with disector sampling. We will review the differences between these estimators and their applications in scientific research. The combination of unbiased disector counting with MOV_N estimation is discussed as a powerful method for generating number-weighted object size distributions. Two primary methods for estimation of local size are shown in fig. 12.1.

These local size estimators are distinguished from one another by two factors, sampling and applications. First, these methods differ in their approaches for sampling the particular objects for local size estimation. The PSI method uses a point grid to sample objects, such as cells and nuclei, for estimation of MOV_V. In terms of applications, the PSI approach is primarily used to make predictions based on preferential sampling from the size distribution of larger objects. An example is given to make estimates of MOV_V of tumor cells from thin tissue sections cut through biopsies of benign and malignant tumors.

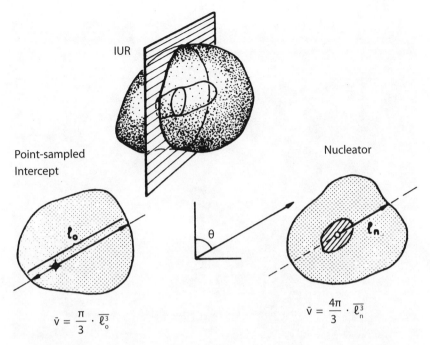

IUR

Point-sampled
Intercept

Nucleator

$$\bar{v} = \frac{\pi}{3} \cdot \overline{\ell_o^3}$$

$$\bar{v} = \frac{4\pi}{3} \cdot \overline{\ell_n^3}$$

FIG. 12.1.   Local Size Estimators. Two estimators of local size, in contrast to region volume.

The value of this PSI approach lies in the ability to predict whether individual tumors are more likely to follow a more acute course, such as metastasize, or to recur as malignancies. In contrast, the nucleator method to estimate MOV_N is carried out in combination with a disector probe on thick sections sampled in a systematic-random manner through an entire reference space.

Point-Sampled Intercepts for Volume-Weighted Mean Object Volume

The PSI approach uses a uniform array of points (point grid) to preferentially sample larger objects for subsequent estimation of MOV_V. In practice, thin sections through the tissue of interest, such as tumor biopsies, are sampled in a systematic-random manner using a point grid. Only the object profiles intersected by points are selected for

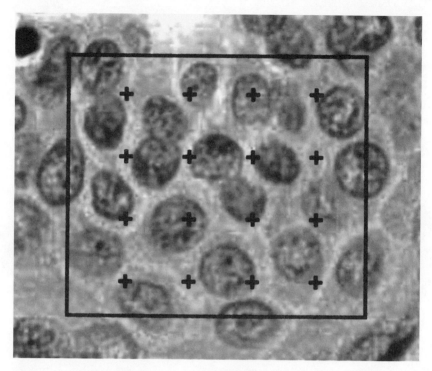

FIG. 12.2. Point Sampling Intercept (PSI) and Melanoma Biopsy Section. An image from a melanoma biopsy overlaid with a PSI grid. Estimates of volume-weighted mean nuclear volume includes volume estimates only from nuclei hit by the point grid; therefore, the estimate is skewed toward larger nuclei.

subsequent estimation of mean object volume. Because larger object profiles have a higher probability of intersecting points, this approach generates a volume-weighted estimate of mean object volume (Gundersen and Jensen, 1985; Gundersen et al., 1988a, 1988b; Sørensen 1989). The PSI approach has been widely applied in studies to quantify volume-weighted mean nuclear volume of cells from benign and malignant tumors. An example of melanoma cells from a tissue biopsy is shown in fig. 12.2.

The PSI approach has been applied in studies to predict pathogenesis of several types of tumors. Certain malignancies, such as malignant melanoma and bladder cancer, have been shown to have strong predictive value for assessing relative metastatic potential (Sørensen,

	≤Median	>Median	Σ
No recurrence (10 years)	9	1	10
Recurrence, No invasion	7	4	11
Invasion	2	12	14
Σ	18	17	35

FIG. 12.3. **Point Sampling Intercept Data for Bladder Tumors.** On volume-weighted mean nuclear volume (mean V_V for nuclei) from $n = 35$ initial benign bladder tumors, expressed as above or below median values, for tumor with different outcomes. Patients with initial benign tumors that showed relatively large nuclear volumes carried a high probability to recur with invasive (malignant) cancer of the bladder. Data from Sørensen and colleagues (see Gundersen et al., 1988a, 1988b).

1989; Ladekarl et al., 1995; Mogensen et al., 1999). For example, fig. 12.3 shows the results for an application of PSI to estimate MOV_V in a study to predict the occurrence of a specific tumor.

In this study, biopsies were obtained from the first benign bladder tumor of cases with three known outcomes: (1) no recurrence, (2) recurrence without invasion (benign bladder tumor), or (3) recurrence with tissue invasion (malignant bladder carcinoma). Tumor sections were stained to reveal the nucleus. A point grid was placed at systematic-random locations. When a point intersected the nucleus of a cell, a line probe was dropped across the nucleus at a random angle relative to a point near the centroid (cell center) and the length of the line measured. This procedure was repeated at different locations through the biopsy, and the mean line length calculated for all the nuclei sampled by the point grid. The resulting mean line length was then cubed, and the product of mean line length cubed and $\pi/3$ calculated to give the MOV_V. These cases were then divided into two groups, those with estimates of MOV_V below and above the median for all cases.

Data collected from the first benign bladder tumor provided predictive information about the probability of each outcome: no recurrence occurred in 90% (9 of 10) of cases with MOV_V less than the median value; recurrence with a second benign bladder tumor occurred in 64% (7 of 11) cases with MOV_V less than the median value; and recurrence with a malignant bladder cancer occurred in 86% (12 of 14) cases with MOV_V greater than the median value. In combination with other clinical stage and histological rating scores for each case, these data provide predictive information to clinicians about the probability of occurrence, as well as whether that occurrence is more or less likely to recur as a malignant carcinoma and therefore require aggressive follow-up treatment (Gundersen et al., 1988a, 1988b).

Nucleator Principle for Number-Weighted Mean Object Volume

A second approach to estimate mean object volume relates to the Euclidean geometry (stereometry) we learned in geometry class. Recall the Euclidean formulas to measure the area of a circle in 2-D ($A = \pi \cdot r^2$) and the volume of a sphere in 3-D [$V = 4/3 \cdot \pi \cdot r^3$). For both of these formulas, the process of data collection is simple: (1) find the exact center of the circle or sphere; (2) measure the radius, r, the distance from the center to the border in any direction; and (3) calculate the area of a circle or volume of a sphere. The only requirement is that the objects in question are either circles or spheres, which raises a problem for applying these formulas in organic stereology, since no biological objects exist as perfect circles in 2-D or spheres in 3-D.

Organic stereology uses variations of these formulas that involve stochastic geometry and probability theory to estimate the volume of arbitrary-shaped object. The Euclidean formulas for the volume of a sphere represent a unique example of the more general concept, the Pappus-Guldinus theorem. The theorem is named after the two mathematicians who independently converged on the same observation some 12 centuries apart: the Greek mathematician Pappus of Alexandria (290–360 AD) and a Swiss professor, Habakkuk (Paul) Guldin (1577–1643). The Pappus-Guldinus theorem states that the volume of a solid object, when rotated at random around a centroid, that is, a point near the center of the object's mass, is proportional to the product of the area of the object and the distance traveled by the centroid.

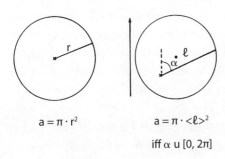

$$a = \pi \cdot r^2 \qquad a = \pi \cdot <\ell>^2$$

$$\text{iff } \alpha \text{ u } [0, 2\pi]$$

Independent of Shape

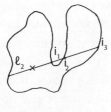

$$\ell_1^2 = \ell_{i_1}^2 = \ell_{i_2}^2 = \ell_{i_3}$$

$$a = \pi \cdot \overline{\ell_j^2}$$

FIG. 12.4. The Nucleator Principle. In contrast to Euclidean-based formulas to measure area of a circle (*left*) or volume of a sphere, the nucleator principle (Pappus-Guldinus theorem) estimates the area of all profiles (*right*) and volume of all objects, regardless of their shape.

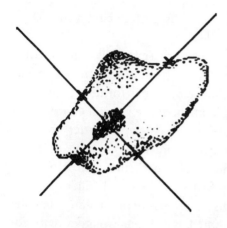

FIG. 12.5. Nucleator Probe Schematic. In practice, the nucleator generates two random lines through the cell (four lines from nucleus to the cell border) for volume estimation. The angle of the first line through the cell is random from 0 to 2π and passes through the cell nucleus; the angle of the second line through the cell is orthogonal to the first.

The nucleator method uses the Pappus-Guldinus theorem to estimate number-weighted mean object volume, MOV_N (Gundersen, 1985), as shown in fig. 12.4. In contrast to the Euclidean formula to calculate the volume of a sphere, the nucleator principle estimates the mean object volume without any assumptions about the shape of the object. For estimating MOV_N the objects of interest are first sampled in an unbiased, number-weighted manner using the disector principle. On isotropic-uniform-random (IUR) or vertical-uniform-random (VUR) sections through the object, a line with a random angle to normal is oriented across a profile through the object, and the distance from the centroid of the object profile to the border measured (fig. 12.5). This process is repeated for objects counted at systematic-random locations through the reference space. To calculate the results, the line lengths for all the objects counted using the number-weighted disector principle are cubed and the mean cubed line length for all the objects determined. The product of the mean cubed line length and $4/3\pi$ is then calculated at the number-weighted mean object volume, MOV_N.

For cases in which IUR sections lead to disruption of tissue landmarks, the tissue containing the objects of interest can be cut into VUR sections and the MOV_N estimated using the nucleator method. With VUR sections, the tissue is rotated around a single vertical axis before cutting sections parallel to the vertical axis. On the resulting

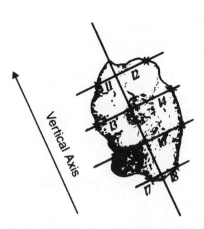

$$V_{obj} = \frac{4}{3}\pi \bar{l^3}$$

where $\bar{l^3} = \dfrac{l_1^3 + l_2^3 + \ldots + l_n^3}{n}$

and n = number of grid lines

FIG. 12.6. The Vertical-Uniform-Random Rotator. The rotator achieves an increase in efficiency over the nucleator by generating more random lines from the nucleus to the cell border, with only negligible increases in time and effort.

sections, the mean length of lines perpendicular to the vertical axis through the objects of interest is directly proportional to the MOV_N of the objects.

Two derivatives of the nucleator method are the rotator (Jensen, 1993) and the optical rotator (Tandrup et al., 1997). The rotator method (fig. 12.6) uses the same Pappus-Guldinus principle as the nucleator but generates unbiased estimates of MOV_N from six to eight line-lengths per object profile. The rotator method is more efficient estimates of MOV_N than the nucleator method because the former requires only slightly more effort; thus, more precision is captured per unit time with the rotator method compared with the nucleator (Jensen and Gundersen, 1993).

A third option to estimate MOV_N using the Pappus-Guldinus theorem, the optical rotator, applies the rotator principle to thick, transparent sections. By focusing through optical planes through the entire section thickness, a larger number of length measurements for estimating MOV_N require minimal extra effort, making this estimator more efficient than either the nucleator or rotator methods. Application of the optical rotator requires computer-assisted hardware-software systems and sufficient section thickness to contain the entire objects of interest.

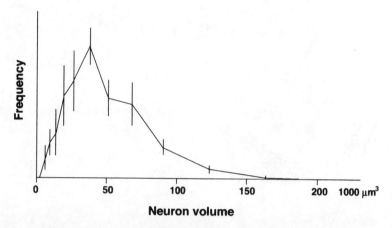

FIG. 12.7. Size Distribution. Number and size of cell (neuron) populations can be efficiently quantified using the unbiased rotator to estimate mean cell volume, in combination with the unbiased disector method to estimate frequency (total number).

Size Distributions

Size distributions are a useful approach to display the number of objects of each size. For example, the combination of the disector principle and the nucleator and rotator methods generates estimate total N and MOV_N, respectively. The combination of these results allows for constructing a size distribution, which provides a useful illustration of changes in both number and size of objects, as shown in fig. 12.7. The area under the curve of a size distribution conveys the total number of cells of all sizes, while the relative skewness of the distribution reveals atrophy (left shift) or hypertrophy (right shift).

13

DO MORE, LESS WELL

- **How Many and How Much Do I Count?**

- **All Variation Considered**

- **Optimization for Maximal Efficiency**

How Many Reference Spaces, Sections, and Locations to Analyze?

How many individuals should be analyzed and with how much sampling effort within each individual? These questions stem from the fact that we cannot analyze all of the individuals in a defined group, and we cannot analyze the entire reference in each of these individuals. Even if we could, the availability of powerful statistical methods for inference testing renders such procedures unnecessary. Instead, design-based stereology relies on the laws of probability to optimize sampling designs for maximal efficiency, an approach termed "Do More, Less Well," by the Swiss stereologist Ewald Weibel.

Sampling and the Laws of Probability

At the most basic level, the goal of organic stereology is to determine the probability that a given geometric probe will intersect the biological object of interest. Once this probability is known, a simple formula converts this probability into a value for the desired first-order stereological parameters of interest, for example, total N.

When a geometric probe, such as a disector for number, point grid for volume, line probe for surface area, or plane for length, is dropped at random into the reference space, two outcomes are possible: either the probe hits the biological object of interest or it does not. If the probe is thrown in the reference space 10 times and hits the object 7 times, then the probability of an intersection is 7 out of 10, or 0.70 (70%). This leads to the obvious question: How many times do I have to drop the geometric probe into the reference space to arrive at a stable estimate of this probability? The answer is the subject of this chapter, but first, consider the following: How many times must I flip a coin to estimate the probability of heads?

Heads or Tails?

Dropping a probe at random into a reference space results in two possible outcomes—either a probe-object intersection occurs ($P = 1$) or it does not occur ($P = 0$)—similar to flipping a coin results in either a head or not a head (tail). In both examples, the outcome is random, and the laws of chance apply.

How many times must a coin be flipped to confirm that the probability of heads is 0.50? We would all agree that 10 flips is not enough. Ten times might give, for example, a result of 7 heads and 3 tails, 2 heads and 8 tails, 4 heads and 6 tails, and so on; finding the true probability of heads clearly requires more work. If not 10 tosses, how many? The answer is that, by about 100 flips, the cumulative number of heads reaches a probability of about 0.50. This does not imply that for any given flip a head is more likely to result than a tail, just that after 100 flips, the cumulative probability of heads is close to known. At this point, the question becomes one of certainty. With 100 flips you might have a result of 48 heads and 52 tails; after 200 flips, the probability will likely converge on 0.50 each for heads and tails.

Consider the situation of dropping a probe at random on tissue containing the biological objects of interest. Provided the sampling scheme and probe geometry are properly designed to ensure that only two outcomes are possible—either the probe hits the biological object or not—then the only factor to influence the probability of a $P = 1$ is the magnitude of the parameter of interest. In this case, we use a similar strategy as the coin toss to estimate the true probability of a probe-object intersection. Only here we need to diverge from the coin toss analogy because the probability of $P = 1$ may be some value other than 0.50. As the probability of $P = 1$ decreases from 0.50, the stringency of sampling to achieve a stable estimate of that probability increases to somewhere between 100 and 200 tosses. The optimal number of tosses will depend on the relative clustering of the objects of interest through the reference space, as discussed in the following section.

Homogeneous versus Heterogeneous Distributions

A common misconception about organic stereology is that the methods somehow apply to quantification of objects with a random distribution through the reference space. Because the sampling and probes are design based to avoid bias, the results will be accurate regardless of the distribution of biological objects of interest. What is true is that the distribution of objects in the tissue dictates the amount of sampling (work) required to obtain a stable estimate of the parameter in a small sample of individuals selected at random from their respective populations.

Because the probability of a probe-object intersection, $P = 1$, is a function of the object's relative distribution through the reference space, different distribution patterns will affect the number of tosses required to make a stable estimate of that probability. As the distribution of objects in the reference space becomes increasingly homogeneous and less clustered, which causes the chance of $P = 1$ to decrease below 0.50, the number of tosses required to make a stable estimate of the true $P = 1$ increases. Take the scenario of a population of objects with a heterogeneous, clustered distribution in which the true probability of $P = 1$ is 0.25. As for the coin toss analogy in which the expected probability of a head is 0.50, 10 tosses is not adequate. About 100 tosses is required to establish the probability of heads 50% of the time; therefore, it follows that the number of tosses will be greater than 100, where $P = 1$ occurs with a probability of less than 0.50. Because $P = 1$ has a 0.25 chance in our population of objects with a heterogeneous, clustered distribution, the optimal number of probe tosses to achieve a stable estimate will approach double the number of tosses required for the coin toss example with probability for $P = 1$ (heads) $= 0.50$; therefore, about 200 tosses of the probe will be sufficient to capture the expected probability of 0.25 for $P = 1$ of objects with a heterogeneous, highly clustered distribution. Similarly, as the clustering of biological objects of interest tends toward a more homogeneous distribution, the number of tosses to capture the probability of an intersection decreases. In this way, the amount of work required to estimate the true probability of a probe-object intersection, and hence, the parameter of interest depends on the relative amount of clustering present for the object of interest in the tissue.

Optimal Sampling Parameters

The optimal parameters for sampling a given reference space may be clear at the start of the study, possibly from previous experience or publications; if not, the expected values for the probability of $P = 1$ are unknown at the start of a study. Options in this case include finding the optimal parameters by a pilot study, which may be effective but require labor. Otherwise, understanding the factors that control the probability of probe-object intersections will help determine the optimal sampling parameters in the most efficient manner.

The relationship between the number of cells counted and estimates of total N is illustrated in fig. 13.1 from the work of Dr. Kebreten F. Manaye at the Howard University School of Medicine in Washington, DC. These data give an empirical foundation for the observation that between 100 and 200 cells counted from the toss of a geometric probe (disector) is sufficient to determine $P = 1$, the probability of a probe-object intersection, as needed to estimate the total number of brain cells (neurons) in a defined reference space.

For each data point on the x-axis, the spacing between disector probe was adjusted to allow between 100 and 1000 probe-object intersections to occur across all 10 sections through a reference space in one human brain. On the basis of the number of cells counted for

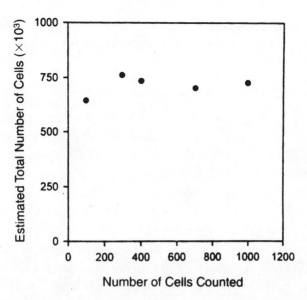

FIG. 13.1. Do More, Less Well. Empirical data showing that counting about 200 objects is sufficient for a stable estimate of total object number using the disector method (unpublished results, with permission from Kebreten Manaye, MD, Howard University College of Medicine, Washington, DC).

each number of repetitions (*x*-axis), the total N was calculated using the optical fractionator method (*y*-axis). The relationship between the *x*- and *y*-axes confirms that the variation between the two possible outcomes ($P = 1$ and $P = 0$) is captured after counting between 100 and 200 cells.

These data illustrate that continuing to toss the probe (disector) into the tissue after counting more than about 200 cells does not appreciably alter the chance of $P = 1$ versus $P = 0$. This empirical test shows that the total N estimate remains stable despite increasing the number of probe tosses to achieve counts of between 200 and 1000 cells. Just as flipping a coin more than 200 times does not appreciably change the cumulative probability of heads, selecting a probe spacing that allows more than 200 disector-cell intersections increases the work to make an estimate of total N, that is, lowers the efficiency, but does not improve the accuracy or precision of the parameter estimate for this individual. Although the data in fig. 13.1 provide information about the optimal number of tosses to obtain a stable estimate of total

N, the same concepts apply to making stable estimates for all first-order parameters using all geometry probes developed for organic stereology.

Within- versus Between-Individual Variability

The data in fig. 13.1 reflects the amount of sampling effort to estimate the probability of a probe-object intersection for a single individual. Using about 200 tosses of the disector probe to count cells in the brain of other individuals from the same group that may have higher or lower values for total N will begin to establish the mean estimate of total N for that group. From these results, we may calculate not only mean (average) values for total N but also standard deviation and standard error of the mean (SEM). However, mean value for any parameter conveys no information about the stringency of sampling used to estimate that parameter. That is, regardless of whether a sampling effort of 10 or 200 tosses were used to estimate total N for each individual, the number of resulting probe-object intersections could be used to calculate total N for each individual, and the mean value calculated for all individuals analyzed.

The coefficient of error, calculated as the SEM divided by the mean (CE = SEM/mean), provides a statistical measure that reflects the stringency of sampling used for estimates of stereology parameters. This leads to the question: What is the optimal CE for a reliable estimate? The decision about the optimal value for the CE requires more knowledge about the degree of variation in total N between individuals within the same group. In the next section, we discuss this biological variability, also known as between-individual variability, which provides the final piece of information to determine the optimal number of tosses of the probe into the tissue for a stable parameter estimate.

Exploratory Tissue Analysis for Organic Stereology

To a certain extent, the level of sampling effort in an organic stereology project depends on the purpose of the study. For example, a study may be designed to estimate a parameter, total N, to an acceptable level of precision; the first step is to cut enough sections through the reference space to capture the majority of between-section variability. Though the exact number will vary for each study, depending on the relative clustering of the objects of interest through the reference space, empirical and theoretical studies have shown that the minimal and maximal number of systematic-random sections through any reference space is from 6 to 12 sections, respectively; that is, about 10 systematic-random sections through a reference space will capture the majority of between-section variability in a parameter estimate (Gundersen and Jensen, 1987; Gundersen et al., 1999). Within this set of about 10 sections, the level of clustering of the objects of interest through the reference space will dictate the stringency of sampling required. Again, as a starting point, about 100 to 200 x-y locations with about 100 probe-object intersections will achieve a stable estimate for each individual. If analysis of $n = 3$ individuals fails to capture 30% of the variation in the parameter for the population, then more individuals must be sampled at random from the population and analyzed.

If, as is the case for most studies to test a specific hypothesis, a stable estimate of the mean value for a parameter is needed to compare two or more groups using powerful methods of inferential statistics, for example, ANOVA, then the critical question becomes, How many individuals must be analyzed, and how many tosses of the sampling probe on tissue sections containing the object of interest to show group differences in the parameter? The next step in an optimal sampling design is to estimate the likely size of the possible effect.

When large differences between groups are present, the effect is "bloody obvious," to borrow a term of exploratory data analysis from the late American statistician John W. Tukey (1915–2000). Inexperienced observers can reliably identify big effects of 50% or more differences in the parameters of interest. The differences between groups are so significant that blinded inspection of biological objects on the tissue sections confirms the presence of obvious group differences in the parameter of interest; if necessary, the principles of organic stereology and statistical testing are used only for attached numbers and degrees of significance for purposes of publication. In terms of statistical power, large effects generally appear in parameters obtained from about $n = 5$ or 6 individuals or less per group. In each of these individuals, relatively light sampling of six to eight systematic-random sections through the reference space and about 100 probe locations across all six to eight sections is sufficient to confirm large effects. When the size of the effect is large and obvious, the most optimal approach may be to sample the two groups (control and treated) with customized sampling schemes. For example, the cases with major loss of cells may be sampled with a rare event protocol, while the cases in the control group sampled at the usual 100–200 x-y positions with a motorized stage.

As the size of the effect diminishes, the number of individuals required to capture the within-group differences increases, along with a higher stringency of sampling with the maximal set of about 10 sections through the reference space, is required to generate a stable estimate. For a moderate-sized effect with group differences in the range of about 20–50%, the number of randomly sampled individuals from each group typically rises from about 5–6 to about 7–10 per group; the optimal number of sections through each reference space increases from 6–8 to 8–12; and the optimal number of probe locations within these sections from about 100 to about 150. For weak effects with small between-group differences of less than 20% for the parameter of interest, confirmation of group differences requires more individuals (up to 10 or more per group) and more sampling effort within these individuals up to about 200 locations across the set of 8–12 sections through the reference space.

As the differences between groups of subjects becomes less pronounced, the need to use a separate sampling scheme for each group lessens. In the final step of exploratory (unblinded) tissue analysis, the most effective way to proceed is to analyze about 10 systematic-random sampled sections from one case to confirm a starting point for setting the optimal sampling parameters. This step is needed to ensure that the expected number of probes fall across all the sections in the study, for example, about 150 for a moderate effect. When the sampling parameters will be constant for all groups, this interprobe distance should be determined by first analyzing cases from the group expected to require the highest level of sampling stringency. That is, in a study to estimate total N after a treatment expected to kill cells in the reference space, an individual from the treated group with the fewest expected number of objects will require the highest level of sampling stringency. In this case, the interprobe distance in the x-y axis should be set based on the analysis of individuals from the treated group. Conversely, if the treatment involves an intervention expected to increase the number of cells in the treated group relative to the control group, then the preferred choice for setting the interprobe distance is an individual from the control group.

The purpose of setting the sampling parameters using an individual that requires the highest sampling stringency is to ensure adequate sampling of all groups. Because the sampling design is optimized for the individual with the lowest probability of a probe-object intersection, the same design will provide sufficient sampling for those individuals in the study with higher probabilities of intersections. Though this approach may result in a slight oversampling of some groups, the sampling design will guarantee that all cases undergo ample sampling using the same sampling design.

Biological Variability

Inspection of fig. 13.1 shows that oversampling beyond the number of tosses required to count about 200 cells simply wastes time and labor resources, without enhancing the parameter estimate. The parameter estimate for any one individual is limited, unless we are comfortable with the assumption that the parameter estimate for this individual is the prototypical example, the statistical average, for all individuals in the group. While such an assumption is convenient, without any information to the contrary, we must also consider the possibility that the total N estimate from this individual represents a statistical outlier and therefore is not a true reflection of the central tendency of total N for the group and population. To address this issue in a systematic manner requires considering the influence of biological factors on the observed variation in a parameter of interest.

The variability between individuals from the same group is biological variability (BV), also known as either *within-group* or *between-individual variation*. Biologically speaking, sample parameters with low levels of within-group variability tend to be more important than sample parameters with greater within-group variability. For instance, consider the human body temperature after a night of sleep. Because the biological processes that support life require a body temperature of about 37°C (±1°C), the body has evolved a variety of homeostatic mechanisms to maintain this narrow normal range. When the first temperature reading after a night of sleep varies outside this narrow range, the body responds with shivering, sweating (evaporation), vascular dilation/contraction, and fever, all designed to normalize the temperature as quickly as possible. If, despite these measures, the body temperature continues to vary outside this range for an extended period, the body's cellular functions will fail. If left unchecked, this failure will lead to death for the individual, which removes an outlier from the temperature distribution for humans, thus moving the central tendency for body temperature in the direction of the previous normal range. In this example, the critical importance of these cellular requirements for the optimal body temperature is reflected in the strong selection pressure to maintain the narrow normal range of body temperature. The length of the radius bone in the forearm is another parameter that varies in the human population. However, this parameter is under less intense selection pressure than body temperature, which frees the parameter to vary across a relatively wide range without negative impact on the survival of individuals. We can conclude that biological parameters associated with relatively narrow normal ranges represent the most important parameters to individuals' survival within that group. An added benefit to the study of parameters with the lowest levels of BV is that less sampling (work) is required to capture the within-group differences in these parameters, that is, fewer individuals to analyze and fewer tosses of the probe to estimate the probability of $P = 1$.

With these theoretical issues in mind, the practical issue remains, how to analyze the optimal number of individuals with the optimal number of probe locations (i.e., tosses of the coin) to estimate $P = 1$, the probability of a probe-object intersection.

The Pilot Study

Exploratory tissue analysis provides an idea about the likely size of the effect and the required information about the optimal interprobe distance to produce a stable estimate for all groups in the study. The next step is to carry out a pilot study in a few individuals from each group in the study. The pilot study purpose is twofold: (1) to collect empirical data on the parameter of interest to optimize the sampling parameters and (2) to identify the amount of statistical power needed to demonstrate group differences, if any.

The typical pilot study begins with two to three individuals from groups most likely to show differences in the parameter of interest. Second, the number of probe locations is intentionally set at a relatively low level, that is, about 100 probe locations. In line with the overall concept of "Do More, Less Well," this strategy avoids wasting unnecessary effort by oversampling within each individual, in the event that the same outcome could be shown using a low level of sampling. That is, the results from the pilot study with a low level of sampling may show statistical differences between groups, in which case the study is complete with no further effort.

If the pilot study fails to reveal a large effect, but shows a trend in that direction, the investigator may decide whether to add more individuals per group and thus increase the chance of revealing statistical differences by improving the statistical power of the study; or, to analyze more locations within the sections from the individuals already processed and analyzed in the pilot study. The consequences of this decision may or may not be critical, depending on the availability of material resources, time, and labor to carry out further studies. Analyzing more individuals carries a strong chance of revealing difference, if they exist. However, if research resources are limiting, consideration should be given to increasing the sampling stringency for the tissue sections already processed to reduce the observed variation between groups to reveal underlying statistical differences.

The latter option conserves resources, and a procedure for partitioning the observed variation is one way to determine whether this option is likely to be fruitful.

Partitioning Observed Variation

Excessive sampling within a single individual is far less efficient than using the same effort to sample a greater number of individuals. This process of focusing work effort on the greatest source of variation has been termed "Do More, Less Well," by Ewald Weibel, a pioneer of modern stereology. To follow this adage, however, some information is required to identify the major sources of variability in the sample parameter estimate.

As detailed throughout this text, organic stereology eliminates or avoids all known sources of bias (systematic error) from estimates of first-order stereology parameters (N, L, S, V). As a result, the variation between groups in final data represents real variation, that is, without error terms arising from the bias associated with false assumptions, faulty models, and inappropriate correction factors.

The workflow for partitioning the observed variance begins with cutting the reference space into about 10 (range 8–12) systematic-random sections, that is, with a random start on the first section through the reference space, with subsequent sections a systematic distance apart. This intensity of between-section sampling ensures that increasing the number of probe locations spaced in a systematic-random manner across these sections is the most efficient method to reveal significant differences between the group mean values for the parameter estimate. Second, a pilot study is carried out using light sampling to estimate the parameter of interest for a few individuals from each group. Finally, the observed variation is partitioned into its sources to provide a rational basis for continuing efforts to reveal statistical differences between groups, if present.

For partitioning the source of variance, two of the three sources of variation are derived from the data, and the third source is calculated as the difference between these values. The first quantity derived from the data is the total observed variation, which is calculated as the coefficient of variation (CV). The CV is calculated for each group as the ratio of the standard deviation (SD) to the mean (CV = SD/mean). To add and subtract variation, which in theory includes the possibility of positive and negative values, the CV is squared to the coefficient of variance (CV^2), which represents the total observed variance in the parameter estimate. The second value derived from the data is the CE, which is squared to the error variance (CE^2), the first of two sources of error that contributes to the total observed variance, as shown in equation (13.1):

$$CV^2 = BV^2 + \text{mean } CE^2 \tag{13.1}$$

where

CV^2 = total observed variance

BV^2 = biological variance

CE^2 = mean sampling variance (mean error variance) = between-section variance + within-section variance

The second contributor to the total observed variance is the biological variation (BV), which squares to the biological variance (BV^2). Rather than a direct derivation from the parameter estimate, the BV^2 is calculated from the other two sources, as shown in equation (13.1). These three sources of error are related in a way that allows for calculating BV^2 through a simple rearrangement ($BV^2 = CV^2 - \text{mean } CE^2$). This process is repeated for both groups analyzed in the pilot study.

Inspection of these sources of variance reveals the total sources of error in the parameter estimate. For example, if the contribution of the mean CE^2 is less than one-half of the BV^2, then reducing the distance between probe locations in both groups, that is, increasing the number of probe locations from 100 to 200, may reduce the variance between mean values for each group to reveal statistical differences at the $P < 0.05$ significance level. If reducing the mean CE^2 by increasing the number of probe locations to 200 fails to reveal significant differences, then two final possibilities exist. The first possibility is that, given the level of statistical power in the study, within-group differences remain too large, and the size of the effect under investigation is too small, to reveal significant differences between the groups. With the sampling stringency already high enough to capture the majority of within-individual variation, the most rational next step is to add additional individuals to each group, that is, process and analyze two or three more individuals per group. Finding the correct number of individuals to add may be done in a step-wise manner, such as two individuals at a time to each group, or by a statistical test known as the *power analysis*. The underlying premise of the power analysis is to solve for the number of individuals from each group required to show a significant difference at $P < 0.05$ level, under the assumption that the standard deviation calculated from the pilot study will remain constant. This assumption leads to a conservative recommendation, that is, tends toward an overestimate, since the standard deviation typically decreases as the number of individuals analyzed within each group increases.

If analysis of more individuals per group to a high sampling strat-

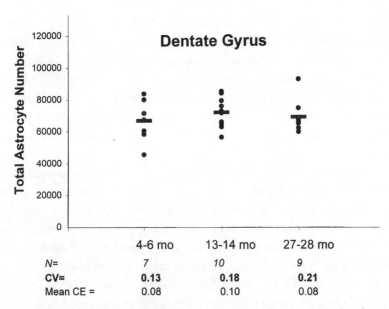

FIG. 13.2. Total *N* Astrocytes in Dentate Gyrus. Data showing age-related variation in optical fractionator counts of total number of cells (astrocytes) in a defined structure (dentate gyrus). The tissue was sampled to a high level of sampling stringency (CE < 0.10) to minimize the impact of sampling error relative to biological variability; CV = standard deviation/mean. CE, coefficient of error; CV, coefficient of variation. Data from Long et al. (1999).

egy (200 probe locations) does not reveal statistically significant difference, or if the power analysis recommends an impractical number of cases to analyze for a statistical difference, then a final possibility is that the groups do not differ from the stereology parameter of interest. How many individuals should be analyzed until the decision is made to accept a conclusion of no differences? The final answer is left to the investigator's discretion and the norms in their particular field, with the following caveat. Increasing statistical power leads to reductions in the degrees of freedom, which, in turn, reduce the level of group differences required to show statistical differences at the level of $P < 0.05$. As the number of individuals per group exceeds $n = 10$, relatively small differences between groups approach the threshold of statistical differences. Thus, the presence of sufficiently high biological differences in a parameter, such that $n = 10$ or more individuals per group must be analyzed to show significant differences, may indicate that the small observed differences between groups are not sufficiently scientifically important to justify further analysis.

One example of this situation is shown in fig. 13.2. The purpose of this cross-sectional study was to test the null hypothesis that age-related differences exist in the total N of cells (astrocytes) in a defined reference space (dentate gyrus) in the brains of male mice. Despite high stringency of sampling ($CE \leq 0.10$) of immunologically stained astrocytes within each dentate gyrus, no age-related differences were found. Since the investigators were primarily interested in the presence of large or moderate effects, no further analysis was carried out. The study was subsequently published without support for the null hypothesis but of strong interest to the bioscience community (Long et al. 1999).

14 . . .

UNCERTAINTY

- • Difference between Bias and Uncertainty
- • Range of Tissue-Processing Sources of Uncertainty
- • Techniques to Avoid Introduction of Nonstereological Uncertainty

Stereological bias refers to systematic error that arises from nonverifiable assumptions, faulty models, and inappropriate correction factors. Previous chapters reviewed the design-based (unbiased) methods for estimating all first-order stereological parameters: volume (V), surface area (S), length (L), and number (N). If all known sources of bias (systematic error) are avoided in these sample estimates, the remaining error in the results will be random—biological variation and error variance. In support of these design-based stereology methods, sampling designs have been developed to partition observed variances and optimize sampling approaches to maximize precision and efficiency. Using the "Do More, Less Well" approach to focus further sampling on the primary sources of error, biological variation is reduced to reveal significant differences, if present, in the most efficient manner.

However, stereological bias is not the only source of nonrandom error that can confound morphometric studies of biological objects. Other sources of nonrandom error arise from the processing protocols used to prepare tissue for stereological analysis. These nonstereological, nonrandom sources of error are known as *uncertainty*. In contrast to stereological bias, which is nonmeasurable and once present in data cannot be removed, sources of uncertainty may be identified and eliminated. Among the numerous potential sources of uncertainty in stereology studies, the most common arise from histological protocols, issues related to microscopy, anatomical identification of reference spaces, and user-directed bias. Though not intended as a complete listing, this discussion of nonstereological bias shows the range of potential uncertainties in stereology studies. Since each investigator is familiar with ambiguities, caveats and concerns in their own studies, each carries the responsibility to identify and correct these sources of uncertainty that may introduce error into the study results.

Penetration Bias

Certain staining protocols achieve greater tissue penetration through sections than others. Histochemical staining methods that use water-soluble dyes, for example, tend to penetrate more easily into tissue than certain other nonwater soluble methods of colorization. If stain penetration through the section is questionable, then quantification of biological objects ostensibly stained by these methods is uncertain. Complete staining of biological objects of interest through the entire reference space must be confirmed before using organic stereology methods to estimate stereological parameters of these objects.

Figure 14.1 shows an example of poor penetration of an immunostain for synaptophysin. Immunostaining was carried out on free-floating sections, and the section was reembedded and resectioned in the orthogonal plane to show the extent of stain penetration. The following section reviews some of the histological factors that could be responsible for poor tissue penetration.

Fixation

Fixation represents a trade-off between the need to maximize tissue preservation while minimizing changes to tissue antigens. The structure of tissue antigens and the influence of binding primary antibodies to specific tissue antigens can be altered by fixation. Fixation is required to preserve tissue morphology and to retain epitopes on an-

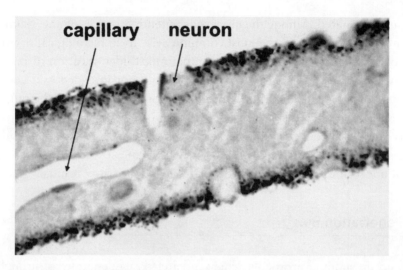

FIG. 14.1. **Penetration Bias.** Tissue through macaque neocortex sectioned in the coronal plane, immunostained free floating with synaptophysin antibody, and then resectioned in an orthogonal plane to show limited antibody penetration (Mouton et al., 1997).

tigens for antibody recognition. The extent of fixation plays a major role in shrinking tissue and identifying biological objects in tissue sections, with some objects not stained in poorly fixed or overfixed tissue.

Formalin, a commercial preparation of formaldehyde, contains 37% formaldehyde in distilled water on a weight:weight basis. Formaldehyde molecules tend to condense spontaneously into long-chain polymers, a reaction that varies as a function of time, temperature, and pressure. This polymerization process reduces the fixation capability of formaldehyde but can be inhibited by alcohols; as a result, many commercial preparations typically contain 10% methanol. A relatively poor fixative, methanol is strongly hydrophobic, which leads to greater tissue shrinkage than aldehydes. Therefore, methanol-containing solutions of formaldehyde can affect antigen:antibody reactions in a manner that ultimately determines the strength of staining. Formaldehyde oxidizes to formic acid, leading to a change in the pH of the stock solution. Because the formic acid concentration varies with time and storage conditions, this effect is low relative to the formaldehyde concentration and, therefore, can be prevented by dilution of formalin in buffers. Aldehyde fixatives also may be combined or mixed with other compounds to enhance their staining characteristics. Acetone and methanol denature and precipitate proteins, while certain antibodies may fail to recognize their cognate antigen after incubation in some fixatives. These factors can potentially affect the intensity of staining biological objects for stereology studies.

Numerous fixatives are available for tissue preservation, including acetone, methanol, and aldehydes, such as formaldehyde, paraformaldehyde, and glutaraldehyde. Aldehydes cross-link proteins in place to prevent leaching from tissue. Chemically reactive aldehyde groups interact with primary amines, such as lysine residues in proteins and thiols, to form cross-links of specific spatial characteristics. As a result, 10% formaldehyde and its 4% paraformaldehyde derivative have become preferred choices for fixation of proteins in the histology laboratory.

Fixation preparation. Proper preparation of fixatives is required for satisfactorily fixation of organic objects for microscopic visualization. Paraformaldehyde powder requires heat and the addition of base to dissolve into solution, and the use of a heating block and a magnetic stirrer minimizes polymerization. For optimal fixation, a 4% solution of paraformaldehyde should be prepared on the day of use as follows: 4 g of paraformaldehyde powder combined with 40 mL of high-quality (distilled or nanopure) water is mixed with two drops of 6N NaOH and placed on a heat block, stirring at 60°C for 5 minutes. Fifty milliliters of cold (refrigerated) Sorensen's phosphate buffer is added and the solution brought to room temperature. As a final step, the pH is adjusted to 7.4 with HCl and the solution brought to a final volume of 100 mL with water.

Fixation duration. Another critical variable for standard tissue processing is the duration of fixation. With a small molecular weight, formaldehyde penetrates tissue rapidly and cross-links formed more slowly. However, actual times vary considerably, for the majority of staining methods, the recommended duration of fixation is minimum (6–8 hours) to retain antigenicity to the greatest extent possible. Variable periods of fixation and using detergent solutions, for example, Triton X, may improve antigen:antibody binding and increase tissue penetration for immunostained sections.

Tissue Preparation

Cryoprotection. Freezing fixed tissue causes the formation of large ice crystals within cells and in the interstitial space. During thawing, ice crystals damage cell membranes, leading to severe artifacts after tissue staining. Typical protocols use sucrose after fixation to prevent ice crystals (cryoprotection) from forming. The tissue is first saturated with sucrose by immersion in a 30% solution (weight:volume) overnight. Initially, the tissue floats in the sucrose solution and then gradually sinks as the osmotic pressure of the tissue and surrounding solution equilibrate. After fixation and cryoprotection, the tissue is frozen in a shallow container containing about 20–30 mL isopentane (2-methyl butane) chilled by dry-ice powder. Once frozen, the tissue is transferred by forceps into a storage vial and quickly placed into the freezer until sectioning.

Tissue sectioning. For frozen and paraffin-embedded sections, sliding and rotary microtomes and cryotomes cut tissue in a systematic manner with a random start at an instrument setting (block advance) of 25–50 μ. For stereology studies, the choice of section thickness depends on the parameter of interest, with thicker sections typically in the range of 40–50 μm for 3-D probes such as the disector, isotropic spheres (space balls), and virtual cycloids. The decision to section in the coronal, sagittal, or horizontal planes depends on the availability of landmarks for identifying anatomical structures. The most efficient choice of sectioning direction is the plane that exposes the most volume of the biological objects of interest on the cut surface of the section. Provided the entire reference space is sampled along each axis, the direction of sectioning does not affect accuracy of stereology estimates. To ensure complete sectioning, reference spaces that exist within larger structures should be "bookended," that is, starting the collection of tissue sections a few sections before and after the reference space, rather than starting the tissue collection by mounting the first section through the reference space on the first glass slide.

Tissue handling. Typically, sections are collected into 12- or 24-well tissue culture dishes (fig. 14.2) containing 2 mL of phosphate-buffered saline, with 100 mM sodium azide to prevent the growth of bacteria, mold, or yeast during protracted storage. After sectioning,

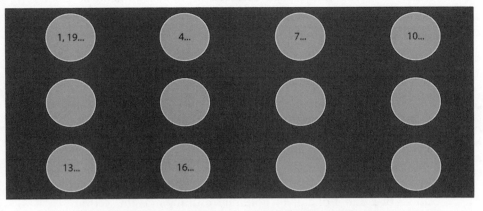

Series #2 Series #3

FIG. 14.2. Multiwell SUR Sampling. Shallow multiwell plates provide a convenient means to organize, store, and retrieve tissue sections cut using the systematic-uniform-random approach. Different staining series are shown in different 12-well plates.

tissue sections can be stored in the cold for long periods (months) without loss of immunostaining. Two 24-well culture dishes side by side at the microtome can receive all sections from an entire mouse brain in the horizontal plane (approximately 1 cm). The first section is placed into column 1, row 1 of the first dish, with subsequent sections placed in order into the following columns of row 1: sections 1–6 in the first culture dish, 7–12 in row 1 of the second dish; section 13 in dish 1, column 1, row 2, and sections 13–24 continue in row 2 of both dishes. Section 25 is placed into dish 1, column 1, row 3, and section 37 is placed into dish 1, column 1, row 4. After completing section 48, section 1 will be present in each well of both culture dishes. This approach maintains the spatial orientation of the brain, allowing the investigator to track sections in order. Section 49 is placed in the same well as section 1, and sections 50–96 continue in order as described, a process that repeats until the brain is completely sectioned with four sections per well. This distance between sections allows sections within a given well to be easily distinguished by eye, using anatomic landmarks, with reference to an atlas if needed. Each of the 12 columns contains 16 sections separated by 300 µm (12 × 25 µm), which can be operationally defined as one set of sections. For each experimental animal, selecting columns can be randomized by rolling dice or with a random number generator. Thus, sample 1 could be column 4, followed by sample 2 as column 7. Sections from each column are stained for a particular marker. If fewer sections are needed, it is convenient to use one-half set of sections by choosing either wells 1 and 3 or 2 and 4 of a given column, resulting in eight sections, each separated by 600 µm. This section spacing is appropriate for spanning the large reference space of large reference spaces, such as hippocampus and cerebral cortex, in rodent brains. Smaller structures or nuclei will require adjusting collection parameters to allow for selecting sections closer together to maintain appropriate systematic-random sampling as described earlier.

Tissue Staining

A variety of staining protocols, including histochemistry, immunohistochemistry, and enzymatic reactions, provide the necessary colorization for organic stereology studies. Though detailed descriptions of these methods are beyond the scope of this text, a few important points related to organic stereology are covered.

Staining intensity. Staining protocols for stereology studies must achieve a minimum threshold for visualization of the biological features of interest, that is, of sufficient intensity to support a binary (yes/no) decision about whether the feature of interest intersects the geometric probe. Staining intensity above the threshold for recognizing positive staining does not introduce variability into the results, in contrast to studies using video densitometry (for review, see Mouton and Gordon, 2010).

Free-floating sections. For stereology studies to estimate stereology parameters, frozen or vibratome tissue sections are often stained free floating, an approach that optimizes the penetration of antibodies and other reagents from both cut surfaces to the center of tissue sections. To avoid confounding artifacts arising from batch effects, samples from all groups in a study are stained simultaneously, that is, using the same protocol with the same solutions and reagents.

Slide-mounted sections. Paraffin-embedded tissue sections are cut using a rotary microtome and then mounted on glass slides for immunostaining. Because penetration is limited to a single surface of tissue sections, a slightly different immunostaining protocol, that is, higher antibody titers and longer incubation times, may be required to produce optimal staining intensity. Slides are typically placed on a heating table to melt paraffin and then placed horizontally on racks in a humid chamber. Staining is carried out in solutions of antibody and other reagents and shifted as needed for washing and color development.

Immunostaining. Kinetics of antibody binding to tissue sections, at concentrations typically used for immunostaining procedures, confirm maximal binding during a 2-hour incubation at room temperature. Because antibodies bind to antigens at two sites, the off-rate for antibodies from antibody:antigen complexes is essentially nonexistent, with virtually no antibody dissociating for more than 72 hours. As a result, tissue incubations with antibodies require sufficient duration to bind for maximal signal, with sections typically held in primary antibody solutions overnight.

When the primary antibody is conjugated directly to a reporter molecule (peroxidase or fluorochrome), binding affinity is reduced and staining less sensitive than other methods. Alternatively, a secondary antibody conjugated to a reporter molecule binds to the primary antibody. Secondary antibodies are generated by injecting a large animal (horse or goat) with the IgG from the animal that produced the primary antibody (mouse or rabbit). Several commercially available kits contain secondary antibodies and color development reagents. For peroxidase-linked methods, these products fall into two levels of amplification. In the first case, secondary antibodies directly conjugated to peroxidase, with one peroxidase molecule bound to each antigen. Second, secondary antibodies can be conjugated to biotin, with tertiary reagents composed of avidin-peroxidase or streptavidin-peroxidase conjugates. Biotin has an extremely high affinity for binding to avidin.

Because avidin contains four potential binding sites for biotin, the avidin-biotin complex (ABC) technique results in formation of large complexes containing numerous avidin, biotin, and peroxidase molecules, resulting in greater sensitivity and amplified color development. A typical protocol is to incubate floating sections with primary antibody overnight at 4°C, followed by sequential incubations in the biotinylated secondary antibody for 2 hours and then in streptavidin-peroxidase for 1 hour. However, for detecting antigens with relatively lower tissue concentrations, longer incubation in primary antibody may be necessary to achieve sufficient penetration for an observable level of tissue staining.

Color development. The peroxidase-diaminobenzidine (DAB) process provides an effective, uniform, and long-lasting colorization reaction to identify specific antigen-antibody reactions. The enzyme peroxidase reduces hydrogen peroxide to water and oxygen, which oxidizes DAB to a dark-brown precipitate. The high-specific activity of peroxidase leads to rapid oxidation of DAB, which swiftly quenches color development. Including various salts (cobalt chloride, nickel ammonium sulfate) will modify color in the final reaction product from brown to black. As a final step, tissue sections should be counterstained with a Nissl stain (cresyl violet or thionine). Counterstaining gives appropriate contrast to recognize cellular details and to locate the correct top and bottom of stained tissue sections.

Among the drawbacks to the DAB colorization method is that toxic reagents necessitate a ventilated hood. Second, some tissues contain sufficient endogenous peroxidase to generate substantial color in the absence of an immunopositive signal. In most organs, for example, the brain, the majority of endogenous peroxidase occurs in red blood cells (RBCs) to detoxify peroxides produced under high oxygen tensions. After the peroxides–DAB reaction, RBCs appear a deep brown color because of their inherent peroxidase activity. Vascular perfusion of the brain with saline before fixation washes out most of this nonselective staining due to RBCs. Endogenous peroxidase activity can be quenched by incubation in a solution containing methanol and hydrogen peroxide. The peroxidase enzyme reduces the hydrogen peroxide, and methanol in the solution precipitates and permanently inactivates the enzyme. For brain tissue, incubation in a solution containing 10% methanol (volume:volume) plus 3% hydrogen peroxide (H_2O_2) in distilled phosphate buffered saline for 30 minutes quenches endogenous peroxidase activity. Tissues with high endogenous peroxidase activity, for example, liver, may require modifying this protocol for optimal results.

Fluorescence. In this staining method, microstructures that express specific proteins bind to antibodies conjugated to fluorescent probes. Rather than by DAB oxidation, however, these microstructures are visible because they contain unstable electrons that absorb light, causing these unstable electrons to jump between high and low energy states. When the electrons drop to the lower valence, the molecule emits light of a particular wavelength, a process known as *fluorescence*. This process is accelerated by high energy, ultraviolet light (UV), with different wavelengths of UV light more effective for different fluorescent probes; to slow emission of light from the fluoroprobe, the tissue stained by immunofluorescence is stored in the dark.

Microscopy-Related Issues

Magnification selection. Data collection for organic stereology is typically carried out at a range of resolutions, from no magnification of gross structures to high magnification of subcellular microstructures. For quantification of structures at no or low magnification, such as volume by the Cavalieri point counting method or surface area with virtual cycloids, the reference area on each section is outlined with no or low-power objectives (2× – 5×). Using the computer-assisted *Stereologer* system, the operator identifies the boundaries of the reference area by "mouse clicking" around the boundary at low magnification.

The optimal objective increases as a function of the need for cellular resolution, with the ideal magnification that allows for the unambiguous identification of the organic structure of interest. A 60× or 63× oil immersion objective with a numerical aperture of 1.4 typically suffices for optimal thin focal-plane scanning in the z-axis required for estimating cell parameters such as number and mean cell volume. Estimating parameters for smaller objects such as synapses, fibers, and capillaries will require magnifications that approach to the resolving power of the light microscope. With the *Stereologer* system, users carry out this recognition using 100× oil immersion with a numerical aperture of 1.4, either by manual "clicking" objects of interest while focusing through the tissue or using automatic analysis based on feature recognition algorithms. Higher resolution for definitive identification of subcellular objects may require enhanced magnification, such as that afforded by high-frequency laser (confocal microscopy) or electron microscopy.

Figure 14.3 illustrates a range of optical artifacts associated with microscopy that may introduce uncertainty into identifying biological objects in tissue. A form of uncertainty (overprojection) arises from the project of 3-D objects in tissue onto a 2-D observation plane (fig. 14.4).

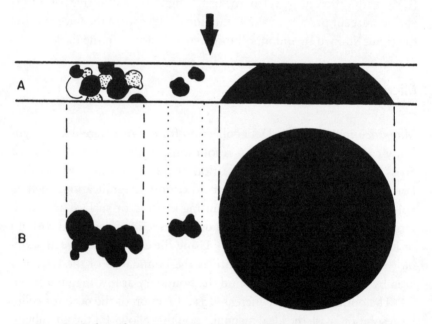

FIG. 14.3. **Optical Artifacts.** A variety of artifacts related to tissue preparation can introduce uncertainty into stereology data. Large arrow indicates tissue section. From left, a group of cells may contain different populations, which, if not properly stained, may appear as the same population; insufficient resolution may fail to distinguish individual cells (masking); and, sections cut too thin can fail to sample large cells.

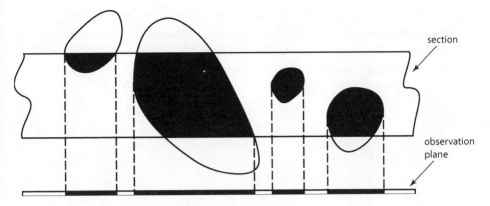

FIG. 14.4. Overprojection (Holmes Effect). Because of the concave shape of objects, the projected size of cells appears larger than the expected (true) value.

Poorly Defined Reference Spaces

Accurate, precise, and efficient quantification of absolute parameters, for example, total *N*, requires an anatomically defined and bounded reference space. Without a clear boundary, subjective delineation of the reference space introduces variation in the data that cannot be accounted for or reproduced. In many cases, poorly defined reference spaces stem from regions defined on the basis of functional characteristics. The advice of anatomists with expertise in the particular structure with consultation of atlases, often assists in locating anatomical borders for estimating absolute parameters using organic stereology.

Summary

Organic stereology assumes that users can identify the biological objects of interest and, in most cases, the anatomical boundaries where these objects occur. A range of protocols, procedures, and practices associated with preparing biological tissue for analysis have the potential to introduce nonrandom sources of error (uncertainty) into this process and indirectly into parameter estimates. The challenge remains for individual investigators to identify the range of these uncertainties that could affect the accuracy of their results and to take the necessary steps to remove them if possible and, if not, to mitigate these uncertainties to the point where they produce only negligible effects on parameter estimates.

15 ...

COMPUTERIZED STEREOLOGY SYSTEMS

- Basic Hardware Requirements for Computerized Stereology

- Operational Features of a Computerized System

- Automatic Approaches for Data Collection and Optimization for Maximal Efficiency

To assist with applying stereology approaches to biological research projects, starting in the mid-1960s, scientists relied on manual (non-computerized) approaches, essentially photomicrographs or microscopic projections onto flat surfaces with geometric probes on transparent sheet overlays. To meet the requirements for systematic-random sampling to capture biological variability with these approaches, a simple stereology study—mean total N of cells in a defined reference space for $n = 16$ individuals in two groups (control and treatment)—required several months to complete. Over the next two decades, these approaches were refined, improved, and applied to an increasing number of research projects in need of reliable stereology parameter estimates from biological tissues.

In the mid-1990s, the Stereology Resource Center (SRC, www.disector.com) established a comprehensive suite of training courses, peer-reviewed publications, scientific consultation services, and computerized stereology systems designed to provide professional support for bioscientific research projects in the international research community. With strong support from a Small Business Innovative Research grant from the National Institutes of Health, the SRC incorporated the principles and practices of organic stereology into the *Stereologer* system, a hardware-software-video microscopy system for computerized analysis of biological microstructures and images (fig. 15.1). Using this system, mean total *N* cell in a reference space for *n* = 16 individuals in two groups can be easily completed in a week by a trained technician using the *Stereologer*. Users now have the option to add image analysis programs that interface closely with the *Stereologer's* camera and motorized stage to ensure seamless image collection, creation of 3-D montages, 3-D reconstruction, segmentation-thresholding, optical densitometry, and other popular features.

This chapter outlines the essential components of a computerized stereology system, along with a list of vendors of hardware components. Because of the changing nature of the electronics, connections, features, and compatibilities of computer systems, camera, and micro-

Computerized Stereology

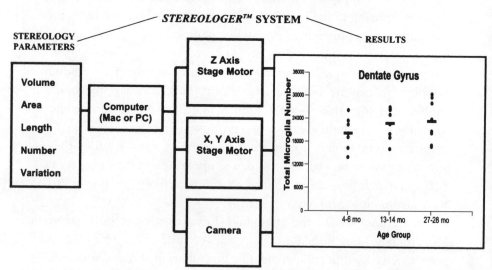

FIG. 15.1. *Stereologer* System. An integrated software-hardware system for high-throughput organic stereology on biological tissue sections or stored images.

scopes, prospective purchasers are strongly urged to contact the SRC for specific recommendations and prices.

Microscope. In theory, all types and brands of upright and inverted microscopes can be equipped for computerized stereology. The primary considerations involve the ability to interface the microscope with the required hardware components under control of the stereology software. The microscope provides magnification and a platform for both the motorized x-y-z stage and illumination for transmitting brightfield and fluorescent images to the computer. Hardware requirements include a camera mount (C-mount) to attach a video, digital, or high-definition camera, and the ability to attach a motorized x-y-z axis stage under control of the stereology software.

The preferred optics for computerized stereology includes at least one high-resolution oil immersion objective, either from 60× to 100×, with the highest numerical aperture possible (NA 1.4). This objective provides the thin focal plane necessary for optical scanning in the z-axis for 3-D probes, such as the disector, isotropic sphere (space balls), and virtual cycloid (fig. 15.2, *lower*), as well as finding the precise top and bottom of tissue sections for determining final, postprocessing section thickness. Otherwise, a range of magnification recommended for computerized stereology include low range (2× to 10×) and middle range (20× to 60× dry) objectives for tissue scanning in the x-y plane and delineation of reference areas on tissue sections, as shown in fig. 15.2 (*upper*).

Computer. The *Stereologer* system supports Macintosh and PC-based computer platforms. Software and the computer's central processing unit (CPU) perform at least three important functions in a computerized stereology system. First, the software/CPU integrates the visual image from the camera with mechanical movement of the stage motor. Second, the software/CPU tracks mouse clicks at probe-object intersections across the probe locations on sections sampled through the reference space. Third, the software/CPU carries out the calculations necessary to convert probe-object intersections into first- and second-order stereology parameters. For systems equipped for automatic stereology, such as Verified Computerized Stereoanalysis (VCS), the software/CPU classifies biological objects into different

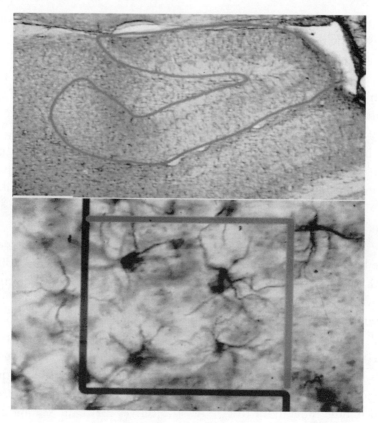

FIG. 15.2. **High- and Low-Magnification Images.** To quantify total number of objects (astrocytes) in an anatomically defined region (dentate gyrus), the reference area on each section is outlined at low magnification (*upper image*), then the number of objects counted using the disector method at high magnification (*lower image*).

types, according to the presence of specific features, for example, color, shape, and texture, and automatically analyzes the first- and second-order stereology parameters. An example of automatic VCS estimation of total amyloid load (Mouton et al., 2009) is shown in fig. 15.3.

Motorized stage motors. Three-axis (x-y-z) motors control and monitor systematic-random sampling of tissue sections under software/CPU control. For computerized stereology, the required accuracy of these motors in the z-axis is high (± 0.10 μ or greater) and relatively lower in the x-y axis (± 1.0 μ).

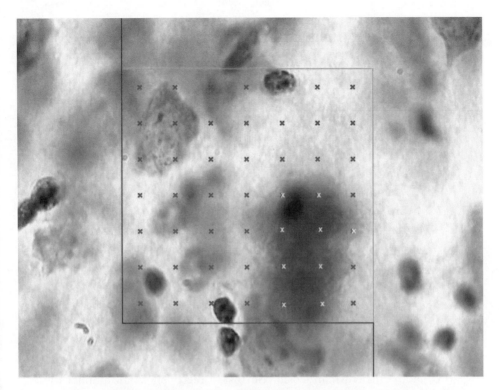

FIG. 15.3. Verified Computerized Stereoanalysis. Using a variety of features associated with different populations of objects, e.g., color, size, shape, and texture, allows for object classification and quantification of first- and second-order stereology parameters.

Camera. Light rays refracted off biological objects in tissue sections carry image information through a series of lenses and mirrors to the photosensitive surface of either a video or digital camera. Images may be analyzed "live" directly from the microscope/camera in real time or as stored images captured off-line, either by the computerized stereology system or from another source, for example, confocal or electron microscope. Although computerized stereology systems support all camera models in theory because of the continually changing connections and electronics by camera manufacturers, prospective users are recommended to consult with vendors of stereology software systems before making a camera purchase.

Operational Features

The overall operations of the *Stereologer*, a parameter-driven computerized stereology system, is outlined next. Before the actual data collection, the system requires a minimal level of input from the user.

Set-up. Data are entered to include study name, data collector, subject ID, reference space of interest, and so forth.

Parameter. Users select the desired parameter of interest (e.g., number, length) and enter an associated label (e.g., "H&E stained cell," "Capillary").

Sampling information. Specific sampling information is provided for each individual case (animal, subject), for example, the total number of sections that contains the reference space and the sampling interval (e.g., every one-sixth section).

Sampling stringency. The user selects the level of sampling stringency for each parameter. For a case with a large number of cells relatively well distributed in a reference space, a low sampling stringency will capture a sufficient number of probe-object intersections (about 100) in an efficient manner. In contrast, a relatively small number of cells in a clustered heterogeneous distribution will require a high sampling stringency to achieve a stable parameter estimate.

Outline reference area. Using a mouse, the user outlines the boundary of the reference area on each section at low magnification (fig. 15.2, *upper*). A point grid is random oriented within each outlined reference area for automatic calculation of region volume. The user switches to high magnification for identification of probe-object intersections (fig. 15.2, *lower*).

Probe-object intersections. On the basis of the parameters selected and labeled by the user in the study set-up, the appropriate probe is generated at the first systematic-random location within the outlined reference area. For manual (nonautomatic) data collection, the user indicates (clicks) for each probe-object intersection, which causes the point-object intersection to change color. The process is repeated for each unique parameter-label combination. For automatic data collection using the VCS approach (fig. 15.4), the software automatically indicates probe-object intersections and presents them to the user for

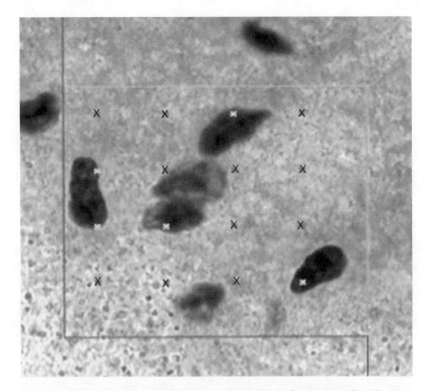

FIG. 15.4. Verified Computerized Stereoanalysis (VCS) for BrDU cells. The size of newly formed brain cells stained with bromo-deoxy-uridine (BrDU) quantified automatically using VCS. BrDU-stained cells at 63× magnification showing probe-feature intersections ("+") using a point grid placed at random over the image. The uppermost left point on the "+" must hit the feature for an intersection to be counted for either manual or VCS methods (from Mouton et al., 2005).

approval. Upon completion of data collection at the first x-y location on the first section through the reference space, the user clicks to proceed to the next x-y location. The user may shift the system into fully automatic mode in which the stage automatically movies to the next x-y location, that is, without approval from the user. For comparison of the relative efficient of VCS versus manual stereology, see Mouton et al. (2005).

Systematic-uniform sampling. In both manual and automatic modes, each sampling location in the x-y axes is selected in an unbiased, systematic-random manner.

Results

All calculations are automatically carried out by the software/CPU. The system software converts the number of probe-object intersections into estimates for each parameter the user selected. The software calculates the various second-order parameters associated with the data, including the total observed variation (CV, coefficient of variation), coefficient of error (CE), and the biological variation (BV). The software analyzes the data for all levels of the sampling hierarchy (between individual, within individual, between section, within section, between probe, within probe) and identifies the levels that make the greatest contribution to the total observed variation. Using the principles outlined in chapter 13, the software provides recommendations to the user for achieving optimizing sampling for maximum efficiency, according to the modern principles of organic stereology. Complete systems with fully integrated hardware-software operation include the following features:

- Complete turnkey Stereologer and Image Analysis system for brightfield and multichannel fluorescent work.
- Capability with both Macintosh and PC computers.
- Full suite of design-based stereology protocols with state-of-the-art user-friendly interface.
- Automatic data management to maximize sampling for optimal efficiency using design-based (unbiased) stereology.
- Full 3-D serial section reconstruction and neuronal tracing capabilities with qualitative morphometric analysis.
- Automatic or semiautomated data collection using Verified Computerized Stereoanalysis.
- Collect, view, zoom, pan, and rotate 3-D image stacks with embedded tracing data.
- Live cellular reconstruction at the microscope in real time or offline from collections of image stacks.
- Automated collection of image stack series for use with design-based stereology protocols.

- Automatic calculation of all stereological results, with full export to Excel and statistical programs.

- Archival collection of large image montages for viewing and printing at low to high magnification.

- Individual profiles for each user saved with settings and program requirements.

- Full imaging analysis capabilities for confocal and electron microscopy.

- Automated functionality for cellular reconstruction.

- Top technical support in the industry from professional stereologists and PhD-level bioscientists.

- On-site installation and training, remote, phone, and e-mail support, full warranty.

As exemplified by the *Stereologer*, the availability of affordable, high-throughput computerized stereology systems, developed with strong support from the U.S. Public Health Service and the National Institutes of Health, allows modern-day bioscientists to apply design-based stereology with high throughput, accuracy, and precision.

Online Demonstrations

For more information and to observe operation of computerized stereology systems, see the three additional online resources available.

1 QuickTime movie on Youtube.com (www.youtube.com/watch?v=80c9fLGfA20).

2 Article at the *Journal of Visualized Experiments* (JoVE) (www.jove.com/index/details.stp?id=1262).

3 Official Web site for the Stereology Resource Center (www.disector.com).

16 . . .

A SURVEY OF TISSUE

- • Stereology as a Survey of Biological Tissue
- • Sampling Hierarchy for Organic Stereology
- • Systematic-Uniform-Random Sampling

The most common reason to use organic stereology in a research project is to test a hypothesis, for example, that an independent variable, such as *treatment* or *disease*, changes some aspect of biological structure, such as the number and volume of cells or surface area of or length of capillaries. The statement of this hypothesis is the most critical step in the experiment and should only be done following careful review of the literature and discussion with colleagues. Once we decide on a specific hypothesis to test that involves a specific change in a morphological structure, the next task is to design a study that uses the methods of organic stereology to test the hypothesis in question.

Organic stereology is like a Gallup poll to discover the truth about a specific question. Suppose, for example, we believe that Ford is the best car company in the United States. To test this hypothesis, we could first design a poll to sample a specific population, such as 100 people outside a factory owned by the Ford Motor Company. Second, we could ask each of these people the same question, "Don't you think that Fords are the best car in the United States?" An overwhelming number of positive responses to this poll would support the hypothesis. In an analogous example from the biosciences, we might hypothesize that treatment with drug A increases in the number of cells (neurons) in an area of the brain that modulates memory. To test this hypothesis, we could sample a population of rats with drug A or placebo and focus our analysis only on an area of the brain that shows a higher density of neurons in the drug-treated group than in the placebo-treated group. Obviously, both of these simple examples represent biased surveys—a study design that favors one outcome over another.

A valid poll to address the question of the best car company in the United States might proceed as follows. First, identify a random sample of registered car owners across the entire United States; and second, avoid asking questions that favor one response over the other. Similarly, to address whether significant changes occur in the brains of rats after treatment with drug A, we might randomly sample an entire brain area that controls memory and use an unbiased probe to quantify the total number of neurons in the brains of drug- and placebo-treated rats.

Sampling Hierarchy

As a first step in testing a scientific hypothesis, we fulfill a critical perquisite for applying powerful methods of statistical analysis when we randomly assign a group of individuals from a single cohort into control and treatment groups. We carry out this random assignment to groups with the understanding that, before any treatment occurs, the true amounts of the particular biological structure of interest varies from one individual to the next in a random manner. In the second step, we expose the individuals in each group to their respective treatments.

Systematic-Random Sampling

From this point on, further sampling within each randomly sampled individual may be carried out using a far more efficient, nonrandom approach—systematic-random sampling—at all lower levels of the sampling hierarchy, which includes reference space, x-y probe locations, and planes in the z-axis. In practice, systematic-random sampling at the next level of the sampling hierarchy, the reference space, is carried out by dividing the reference space into a number of sections, and then identifying the distance between samples that will result in collecting the desired number of samples. For example, suppose that a particular reference space is sectioned at an instrument setting (block advance) of 40 μm. Serial sectioning through the entire reference space at this thickness results in a total of 60 sections. To capture the between-section variation in the parameter of interest, the decision is made to sample a total of 10 sections in a systematic-random manner from this set of 60 sections that includes the entire reference space; thus, the section sampling fraction (ssf) will be one-sixth. With a random start in the first six sections (1 through 6), the first section is sampled, for example, section 5, and then every 6th subsequent section sampled in a systematic manner. In this example, the sampled sections would be the following series: 5th, 11th, 17th, 23rd, 29th, 35th, 41st, 47th, 53rd, and 60th. Because the first section is always taken with a random start, there is a strong probability that the sample of sections through the reference space of the next individual will sample a different portion of the reference space. This procedure allows all sections through the reference space of each individual to have an equal probability of inclusion in the sample, thus ensuring an unbiased sample of tissue through the reference space of each individual for stereological analysis.

The same systematic-random approach is used to carry out unbiased sampling on each sampled section, the next level of the hierarchy. As shown in fig. 16.1, the location of the first probe in the x-y plane of the first sampled section is selected at random in the x-y interval, with subsequent probe locations spaced a systematic distance apart.*

Thus, systematic-random sampling achieves the goal of a well-designed survey in which each part of the reference space carries an equal probability of inclusion in the sample. In some cases, additional criteria may be used to enhance throughput efficiency by sampling through nonuniform sampling of biological objects based on object size, for example, point-sampled intersects to estimate volume-weighted

*This pattern is called a *bostrophedon*, literally ox-turning, to refer to the movement of an ox while plowing a field, from the Greek *bous* (ox) and *strophe* (turning).

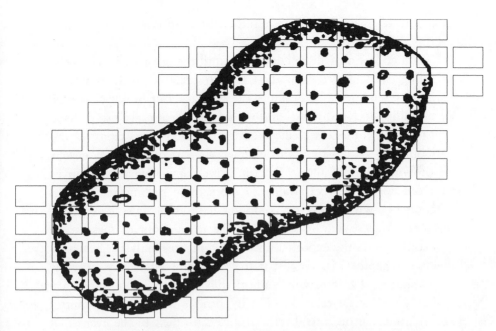

FIG. 16.1. Systematic Uniform Random (SUR) Sampling. Unbiased sampling with a random start at the first location, with subsequent locations a systematic-uniform distance apart. Ideally, about 100 to 200 sampling locations are distributed within in the *x-y* planes of 8–12 sections cut through the reference space in an SUR manner, i.e., first random, subsequent systematic-uniform distance apart.

mean object volume, or the proportionator to estimate total object number for highly clustered object distributions.

Once the unbiased sampling for our stereology probe is complete, the investigator will select the appropriate unbiased probe, similar to the use of an appropriately unbiased questionnaire, to estimate the probability of probe-object intersections (chapter 5). Options for this probe include point grid to estimate a defined region volume, total *V*, using the Cavalieri point counting method (chapter 7); the surface area, total *S*, of cells using the virtual cycloid method (chapter 8); length, total *L*, of capillaries or processes using the space balls probe (chapter 8); the total number of cells, total *N*, using the optical fractionator method (chapter 10); and mean cell volume using the rotator method (chapter 12). To avoid the reference trap (chapter 10), the dependent variables quantified for each individual in this study will be expressed as total parameters rather than ratio estimators, that is, total N, not N_V.

Data collection will be continued until the error variance due to sampling, as estimated using the CE, is low relative to the biological variability (chapter 13). Because a certain level of biological differences within each individual and within each group, we use the concept of "Do More, Less Well" to design our studies to "capture" this variability in the most efficient manner. As part of our study design, all known sources of uncertainty (nonstereological bias) are removed to avoid adding this variability to the sample parameter estimates (chapter 14). With access to a computerized stereology system such as the *Stereologer* (www.disector.com), the throughput of these study will converge on the expected time of 1 hour for analysis of each individual (chapter 15). Finally, if the treatment effect is sufficiently large, statistical testing will confirm that changes in the biological structure of interest related to our hypothesis will cause the control and treatment groups to diverge from each other. In the event that analysis with sufficiently high statistical power fails to reveal differences in support of the initial hypothesis, then one possibility is to report no significant effects of the independent variable.

Like any poll intended to discover the truth rather than confirm a particular viewpoint, the principles of organic stereology are designed to avoid all known sources of bias. Though perhaps not always returning the desired outcome, the applications of organic stereology to bioscientific research will allow science to continue growing at an optimal rate by revealing the truth about natural phenomena. As research bioscientists, our commitment to this task is crucial, not only for the benefit of our individual research programs but also for the sake of science progress into the new millennium.

17 ...

PEER REVIEW CONSIDERATIONS

- Historical Context for Morphometric Approaches
- Minimum Methodology for Publication of Organic Stereology Studies
- Considerations of Peer-Reviewed Funding Issues

A Bit of Context

In the years leading up to the advent of organic stereology, the application of design-based stereology to analyze organic objects, a primary approach for morphometric analysis of biological tissue, included expert-based approaches. The expert-based methods included rating scales (e.g., $+1$ to $+5$) that, while generally reproducible by the same expert, generally varied to a great extent between experts, thus high intra-rater and low inter-rater reliability. This trend likely arose from the wide differences in the training, experience, expertise, and interest among experts, even for those within the same field. One other issue with expert-based morphological analysis was the general paucity of experts willing to collaborate on the large and growing number of studies that required accurate, precise, and efficient morphometric analyses of biological tissue.

To address this demand, in the 1960s, computer scientists led the development of semiautomatic image analysis systems to quantify biological tissue based on the assumption- and model-based methods of Euclidean geometry. Though designed to give results with high efficiency without the input of experts, these computer-based approaches produced a wide range of results with questionable veracity.

Starting in the 1970s, under the influence of Hans Gundersen and colleagues affiliated with the International Society for Stereology (ISS), a new type of thinking emerged through the problems associated with morphometric analysis of tissue. Gundersen and colleagues, including Luis Cruz-Orive (Spain), Adrian Baddeley (Australia), Eva Jensen (Denmark), and Arun Gokhale (United States), systematically identified the wide range of biases and uncertainties in existing approaches, the so-called methods of biased stereology. By using un-biased sampling and parameter estimation based on stochastic geometry and probability theory, this faction categorically rejected all assumptions, models, and correction factors based on Euclidean geometry. They developed a new wave of thinking that shifted the paradigm and ushered in the modern era of design-based stereology. Organic stereology combines the theoretical principles of modern, design-based (un-biased) stereology with the practical caveats and considerations associated with the application of those concepts to organic tissue.

Reviewer Considerations

In 1996, the *Journal of Comparative Neurology*, one of the oldest and highly regarded journals in the neurosciences, published a review by Coggeshall and Lekan of the pros and cons of design-based stereology to estimate numbers of neurons and synapses in defined brain regions. As part of this review, the authors surveyed a sample of neuroscience journals during a 3-month period and arrived at the surprising conclusion that about one-third of published studies in the neurosciences focus on at least one of the first-order stereology parameters. To accompany this review, the journal's editor-in-chief, Clifford Saper, MD, professor of neurology at the Harvard Medical School published an unusual editorial (Saper, 1996) in which he shared his editorial preference for studies that use state-of-the-art approaches of modern stereology over model- and assumption-based methods.

> Stereologically based unbiased estimates are always preferable for establishing absolute counts or densities of structures in tissue sections.

We expect that any papers that use simple profile counts, or assumption-based correction factors, will produce adequate justification for these methods. Referees are urged to . . . insist on unbiased counts when [this justification] is not adequate. (Saper, 1996)

Saper's review outlined the journal's new policy that required design-based stereology, where appropriate, and expected authors who submit manuscripts with morphometric data collected by other methodologies to justify those methods in relation to unbiased stereology. Although several other journals had similar, though less explicit, policies in place at that time, Saper's editorial resonated throughout the international neuroscience community about the value of new stereological approaches. Since then, other peer-reviewed bioscience journals and other peer-review groups, including reviewers of applications for research funding and regulatory approval, have followed Saper's initiative.

Resistance to the "new stereology" came from established biologists who for years, and in some cases decades, had relied on the older, assumption- and model-based approaches for studies that involve quantifying neurological structures. As in many instances, the inertia of tradition opposed the broad acceptance of organic stereology by the international community of bioscience researchers. Authors of studies using older assumption- and model-based approaches did not wish to abandon these methods in favor of unbiased stereology. Other critics expressed views that this approach was too sweeping and did not follow the time-honored tradition of step-by-step progress built on the existing body of knowledge. Unbiased stereology proponents under the multidisciplinary auspices of the ISS argued that the existing approaches based on such assumptions as, "assume a cell is a sphere," simply did not apply to organic structure. Because the systematic error introduced by these approaches could not be quantified or removed, these Euclidean-based methods should be rejected in their entirety, rather than replaced in a piecemeal fashion. In some cases, scientists new to the field of unbiased stereology misconstrued the term *bias*, which, like the term *theory*, has different connotations in lay and scientific settings. Still other biologists objected to the staunch support for the new approaches, which some perceived as "religious fervor." Since the start of the twenty-first century, these objections have diminished for the most part or at least moderated from a roar to a murmur, as evidenced by the broad acceptance of design-based stereology by many peer reviewers, project managers, funding agencies, and government organizations that support and regulate bioscientific research. The next section discusses the types of information that, when included in peer-reviewed publications, tends to convey to reviewers that a particular study possesses the methodological foundations of state-of-the-art organic stereology.

The Value of Pilot Studies

Despite a general agreement among bioscientists that modern approaches of organic stereology represent state-of-the-art methodology to analyze biological tissue, to date there is no general consensus about what constitutes minimal information to publish these studies. However, a minimum set of information exists to communicate the application of design-based principles and practices (as summarized in table 17.1). While no guarantees exist that manuscripts submitted to peer-reviewed journals will be accepted on the basis of any specific information, including information in table 17.1, at least these details provide reviewers with insight into the author's strong understanding of the stereology approaches in the study.

Similarly, though many peer reviewers of funding applications

---

TABLE 17.1. **Minimal information for publications with organic stereology**

Definition of clear, unambiguous reference spaces

Calculate and report reference volumes (V_{ref})

Sampling design, e.g., systematic sampling with a random start

What are the sampling fractions, e.g., every fifth section?

What was counting item (Chat-pos neuronal cell bodies, nucleoli, TH-pos fibers)?

Specific stereological probe (optical fractionator, rotator)

How many animals (age, gender), sections, and actual cells counted per reference space, e.g., ΣQ^- between 150 and 200 cells?

Descriptive statistics: means, variation (SD, SEM); and variance analysis (CV, mean CE), with extra credit for calculation of biological variance (BV) to verify optimal sampling stringency

---

CE, coefficient of error; CV, coefficient of variation; SD, standard deviation; SEM, standard error of the mean.

agree about the value of organic stereology approaches over assumption- and model-based approaches, there is no explicit agreement among scientists who review proposals for extramural support of bioscientific research as to what constitutes evidence of good morphometry. There is evidence, however, that reviewers of funding proposals carefully assess the scientific validity of the methodologies in these applications. According to the Center for Scientific Review, the organization charged with evaluating research grant applications funded by the National Institutes of Health, most unfunded applications suffer from one or more of the problems listed in table 17.2. Notably, three of the nine examples (see statements in boldface) could be avoided with a pilot study using organic stereology.

Lack of experience in essential methodology. Through a pilot study approach outlined in chapter 13, and by providing the information in table 17.1, applicants will demonstrate to reviewers their experience with methodologies for testing hypotheses using state-of-the-art organic stereology methods.

Lack of sufficient experimental detail. By including the information described above in table 17.1, applicants provide details about the experimental details of their planned studies using organic stereology.

Unrealistic large amount of work. Often bioscientists without pri-

TABLE 17.2. **Common basis for rejection of applications for research funding**

Lack of new or original ideas

Absence of acceptable scientific rationale

Lack of experience in essential methodology

Uncritical approach

Diffuse, superficial, or unfocused research plan

Lack of sufficient experimental detail

Lack of knowledge of published relevant work

Unrealistic, large amount of work

Uncertainty concerning future direction

Source: Center for Scientific Review, National Institutes of Health.
Note: Problems that could be avoided with a pilot study using organic stereology are in boldface.

mary experience with organic stereology either overestimate or underestimate the amount of work required to test the hypotheses in their proposed studies. By completing a pilot study and providing the data from this study, as given in table 17.1, applicants will gain the primary experience with the methodology necessary to make these predictions in a realistic manner.

In summary, for peer-reviewed journals and applications for extramural funding, a pilot study in a small group ($n < 5$) of individuals from the control and treatment groups can be generated in support of statistical trends or differences and thus support a hypothesis related to morphological changes. However, pilot data also provide experience and results to address potential concerns of peer reviewers. A pilot study reveals valuable information about time, labor, and material resources needed to complete proposed studies.

APPENDIX: Conceptual Framework for Organic Stereology

The application of design-based stereology to organic tissue:

- Currently is a state-of-the-art approach for morphometric analysis of organic tissue
- Increasingly is expected by journal editors and grant reviewers
- Encompasses five decades of developing theoretical and practical aspects
- Encompasses all anatomically well-defined populations of biological structures
- Demonstrates that the tissue processing requirement differs from earlier assumption- and model-based methods
- Avoids all known sources of theoretical bias, e.g., "assumes a cell is a sphere," and methodological uncertainty
- Avoids inappropriate, i.e., Euclidean-based, correction formulas that do not apply to organic tissue
- Focuses on reliable estimates of total (absolute) parameters (total cell number), rather than ratios such as density (number per unit volume)
- Does not require computerized hardware-software systems, but . . .
- Allows automatic data collection systems that cause dramatic increases in throughput, without a loss of accuracy, compared with manual approaches

- Uses a small sample of individuals to estimate first-order stereological parameters of biological interest, e.g., V, S, L, N, and their variability
- Uses unbiased and high-efficient systematic-random sampling
- Creates efficiency based on true biological variability
- Presents estimates that refer to functionally defined, unambiguously identifiable reference spaces
- Optimizes sampling for maximal efficiency ("Do More, Less Well") using estimates of sampling error and biological variability
- Relates to all biological objects that are reliably visualized in tissue
- Avoids all known tissue processing artifacts
- Is founded on stochastic geometry and probability theory; however, . . .
- Does not require an advanced mathematical background for users
- Allows statistical power to accumulate for similar parameters in different cohorts

GLOSSARY

Accurate   Lack of systematic error (bias).

Anisotropy   A preferred spatial orientation in 3-D.

Arbitrary-shape   Variable structure in 3-D; nonclassical.

Area   First-order stereological parameter for planar (2-D) structures; estimated without bias using a probe with at least one dimension (1-D line).

Area fraction   Object profile area relative to a sampled area ($A_{object}/A_{sampled}$).

Area sampling fraction (ASF)   Frame area divided by area of the x-y step.

Ascertainment bias   Systematic error arising from biased sampling of a target population.

Assumption-based morphometry   Morphometry based on nonverifiable assumptions or faulty models, e.g., organic objects quantified by methods of Euclidean geometry.

Between-section variation   Error in a parameter estimate that arises from section sampling.

Bias   Systematic nonrandom error; with increased sampling causes data to deviate from the expected (true) value.

Biological variance (BV^2)   Squared biological variation; contribution of interindividual (within-group) differences to total observed variance.

Biological variation (BV)   Random error caused by individual differences.

Cardinality   Structural property of number, the first-order, zero-dimensional stereological parameter.

Coefficient of error (CE)   Error arising from within-individual sources of variation; composed of between-section error and the nugget effect.

Coefficient of variation (CV)   Total observed variation in a sample parameter estimate; includes biological variation and sampling error.

Correction factor   Assumption- and model-based component of a morphometric equation.

Cycloid   A sine-weighted line; used in organic stereology as probe to estimate surface area and length from vertical sections and vertical slices, respectively.

Delesse principle   Concept developed by August Delesse (1847), which states for a random section through a 3-D volume containing objects of interest, the area fraction of the object profiles per unit area sampled (A_A) is equivalent to the volume fraction of the objects per unit volume (V_V); that is, $A_A = V_V$.

Density   A ratio estimator with a first-order stereological parameter (numerator) per unit area or volume (denominator), e.g., length density (L_V).

Dependent (systematic, nonrandom) sampling   Scheme in which adjacent structures are sampled in serial order.

Design-based stereology   Morphometric approach to estimate first- and second-order stereology parameters using sampling and probes based on probability theory and stochastic geometry.

Digitization   Image analysis process in which a light image is converted to picture elements (pixels).

Disector   A virtual 3-D probe for unbiased estimation of number.

Disector principle   Unbiased method for counting the number of object "tops" in a known volume of tissue.

Distribution   The spread of values across a target population.

Do More, Less Well   Strategy to optimize morphometry for maximum efficiency in which sampling effort reduces total observed variance by focusing sampling (work) on the relatively largest sources of error.

Efficiency   Precision units per unit of time.

Embedding   A process that involves placement of tissue in a hard matrix for microtome sectioning.

Error   Variation in a parameter estimate.

Error analysis   Post hoc process in which observed error is partitioned into sources.

Error variance   The random within-individual error in a sample estimate; noise.

Estimate   The approximate determination of a parameter in a small random sample from the population.

Euclidean geometry   Field of geometry (stereometry) concerned with morphometry of classically shaped structures, e.g., spheres, cylinders, tetrahedrons.

Exclusion lines   On an unbiased counting frame, lower and left lines where no probe-object intersections are counted.

Expected value   True value of a parameter at the population level.

Exploratory tissue analysis   Qualitative examination of tissue sections for possible within-section effects (clustering) and between-group (treatment) effects; pilot studies to optimize sampling before blind analysis using organic stereology.

First-order stereological parameters Volume (V), surface area (S), length (L), number (N).

Fixation Process of treating organic tissue with chemical reagents to stabilize proteins for histological analysis.

Fractionator A three-level sampling procedure in which total parameters are estimated from measurements in a known fraction of the total reference space.

Guard zone (guard volume) Volume of tissue between probe (disector) and cut surfaces of tissue sections where no data collected; functions to avoid artifacts (e.g., lost caps) caused by the microtome knife.

Inclusion lines On an unbiased counting frame, upper and right lines where probe-object intersections are counted.

Individual differences Expected (true) variability in a parameter between individuals in a defined group.

Interpretation bias Nonrandom error (bias) in a parameter estimate contributed by experimenter-based expectations for a particular outcome.

Isotropic-uniform-random (IUR) sections Method for overcoming anisotropy by sampling organic objects in which all orientations in 3-D have equal weight.

Isotropy Lack of a preferred spatial orientation.

Length First-order stereological parameter for lineal (1-D) structures; estimated without bias using a probe with at least two dimensions (e.g., 2-D plane).

Line probe Uniform grid of test lines (1-D).

Mean (average) Central tendency of a normal distribution; requires at least one degree of freedom, i.e., data from $n \geq 2$ individuals.

Mean object volume (MOV) Local estimate of object size for a sample of organic objects.

Method error Total amount of random within-individual sample error; estimated by coefficient of error (CE).

Methodological bias Uncertainty that arises from techniques required to estimate sample parameter, e.g., recognition bias, staining artifacts, penetration bias.

Microtome Mechanical device for cutting tissue into systematic sections with uniform thickness.

Morphometry Generic term for the cumulative methodologies used to quantify organic tissue in bioscientific research.

Number First-order stereological parameter for number (0-D) of objects; estimated without bias using a probe with at least three dimensions (e.g., 3-D disector).

Number-weighted parameter   Parameter estimated with a sampling approach in which all objects have equal probability of being sampled, i.e., without bias from size, shape, orientation.

Object   Level of biological structure identified at high magnification; also known as *microstructure*.

Optical disector   A design-based approach for estimating total N using thin focal-plane scanning with an unbiased counting frame through a known distance in the z-axis of tissue sections.

Optical fractionator   An unbiased sampling for estimating total N based on counting objects using the disector principle in a known fraction of a defined reference volume.

Optimization for maximum efficiency   Post hoc strategy to direct sampling effort at the sources of error in a parameter estimate that contribute heaviest to the total observed variability.

Orientation bias   Systematic error arising from a nonrandom 3-D orientation.

Overprojection (Holmes effect)   Systematic error arising from projecting concave objects through transparent embedding material; reduced by minimization of section thickness.

Pappas-Guldinus theorem   Volume of a solid object, when rotated at random around a centroid, i.e., point near the center of the object's mass, is proportional to the product of the area of the object and the distance traveled by the centroid. Named after Pappus of Alexandria (AD 290–360) and the Swiss mathematician Habakkuk (Paul) Guldin (1577–1643).

Paraffin-embedding   A tissue-processing approach that uses paraffin to physically stabilize tissue for thin sectioning with a rotary microtome in preparation for histological staining.

Parameter   A quantity at the population level.

Pattern recognition   Technique of computerized image analysis for automatic identification of objects based on defined features.

Physical disector   Application of the disector principle to estimate total N using pairs of adjacent physical tissue sections.

Pilot study   Exploratory tissue analysis to make qualitative estimates of within-section, between-section, between-individual, and between-group variability.

Point counting   Approach for area and volume estimation using a zero-dimension (0-D) point grid with known grid constant (area per point).

Population   A collection of objects with at least one common attribute.

Precision   Observed variation (error) in individual sample estimate that arises from sampling stringency (work).

Probe   A geometric concept with known properties and dimensions, e.g., line-grid (1-D) with 1000 μ of total length, used for sampling and estimating stereological parameters.

Profile   2-D cross-section of a 3-D object caused by sampling with a 2-D plane (e.g., knife blade).

Projection bias   Uncertainty that arises from projecting 3-D structures using light microscopy.

Random sampling   Sampling approach that uses random sampling; may be dependent or independent.

Recognition bias   Systematic error that arises from the inability to unambiguously identify objects of interest.

Reference space   Anatomically bounded volume of biological interest.

Reference trap   Systematic error that arises from use of density estimators to measure absolute quantities.

Region   Level of organic structure identified at low magnification.

Region of interest   In image analysis, the boundaries of structure to be analyzed; may be anatomically bounded or arbitrary.

Rotator   An application of the Pappas-Guldinus theorem for efficient estimation of mean object volume.

Sample   A subset of the population (noun); to obtain a subset of the population (verb).

Sampling error (within-sample error)   Random variation related to the sampling stringency within an individual.

Sampling interval (k)   Period between successive samples through a reference space, e.g., every kth section.

Second-order stereological parameter   Variability of volume, surface area, length, and number.

Section sampling fraction (SSF)   Number of sections analyzed divided by the total number of sections through the reference space.

Shape bias   Systematic error caused by variability in shape of nonclassical objects.

Size bias   Systematic error attributed to variability in object size.

Standard error of the mean (SEM)   Standard deviation divided by the square root of the number of individuals in the sample.

Stereologer   An integrated hardware-software system for design-based analysis of biological structure using organic stereology.

Stereology   The analysis of objects (Greek).

Stochastic geometry   The analysis of variable-shaped objects using probability-based approaches.

Systematic error   Nonrandom variation (bias) in sample parameters from expected values; once present, it cannot be measured or removed.

Systematic-uniform-random (SUR)   Sampling approach in which a reference space is sampled with a random start at the first location and thereafter in a systematic-uniform manner.

Thickness sampling fraction (TSF)   Disector height divided by the final post-processing section thickness.

Unbiased counting frame   An area frame with inclusion and exclusion lines to ensure that the number of objects is sampled with equal probability.

Unbiased method   Stereology method in which increased sampling causes a sample parameter to converge on the expected value for a parameter.

Underprojection   Lack of information caused by embedding media opaque relative to the object of interest.

Variation (variability)   Error; may be systematic (bias) or nonsystematic (random).

Verified Computerized Stereoanalysis (VCS)   An innovative hardware-software approach for automatic analysis of organic tissue in stained sections.

Vertical axis   Axis of rotation in a defined direction.

Vertical section   A random plane parallel to the axis of rotation.

Vertical-uniform-random   Sampling method to avoid systematic error when estimating surface area and length of anisotropic structures in conjunction with cycloid line probes.

Virtual cycloids   Unbiased method for surface and surface-density estimation on sections but at any convenient orientation.

Virtual isotropic spheres (space balls)   Unbiased for length and length density estimation using a sphere probe on tissue cut at any convenient orientation.

Virtual sectioning   Optical sectioning technique that uses a stack of parallel focal planes to probe objects.

Volume. A first-order stereological parameter for 3-D structures; estimated without bias using point (0-D) probe.

Volume-weighted sample. A subset of objects or regions sampled on the basis of volume.

Within-sample error   Random variation within an individual; measured by CE.

Within-section variation   Random variation within a section caused by the smallest unit of variation of a region or object.

BIBLIOGRAPHY

Abercrombie, M. (1946). Estimation of nuclear population from micotome sections. *Anatomical Record* 94:239–247.

Acer, N., Sahin, B., Usanmaz, M., Tatoğlu, H., and Irmak, Z. (2008). Comparison of point counting and planimetry methods for the assessment of cerebellar volume in human using magnetic resonance imaging: A stereological study. *Surgical and Radiologic Anatomy* 30:335–339.

Baddeley, A., Gundersen, H. J. G., and Cruz-Orive, L. (1986). Estimation of surface area from vertical sections. *Journal of Microscopy* 142:259–276.

Baddeley, A., and Jensen, E. V. (2004). *Stereology for Statisticians.* Monographs on Statistics and Applied Probability, 1st ed. Chapman & Hall/CRC, Boca Raton, FL.

Berlin, M., and Wallace, R. B. (1976). Aging and the central nervous system. *Experimental Aging Research* 2:125–164.

Berman, R. F., Pessah, I. N., Mouton, P. R., Mav, D., and Harry, G. J. (2008). Low level neonatal thimerosal exposure: Further evaluation of altered neurotoxic potential in SJL mice. *Toxicological Sciences* 101:294–309.

Berman, R. F., Pessah, I. N., Mouton, P. R., Mav, D., and Harry, G. J. (2008). Modeling neonatal thimerosal exposure in mice. *Toxicological Sciences* 103:416.

Boonstra, H., Oosterhuis, J. W., Oosterhuis, A. M., and Fleuren, G. J. (1983). Cervical tissue shrinkage by formaldehyde fixation, paraffin wax embedding, section cutting and mounting. *Virchows Archiv (Pathol Anat)* 402:195–205.

Braendgaard, H., and Gundersen, H. J. (1986). The impact of recent stereological advances on quantitative studies of the nervous system. *Journal of Neuroscience Methods* 18:39–78.

Brody, H. (1955). Organization of the cerebral cortex. III. A study of aging in the human cerebral cortex. *Journal of Comparative Neurology* 102:511–516.

Buffon G. L. (1977). Compte de essai d'arithmétique morale. *Supplemental histoire naturelle*, vol. 4. Imprimarie Royale, Paris.

Burke, M., Zangenehpour, S., Mouton, P. R., and Ptito, M. (2009). Knowing what counts: Unbiased stereology in the non-human primate brain. *Journal of Visualized Experiments*, no. 27. Video, www.jove.com/index/details.stp?id= 1262.

Cajori, F. (1923). The history of notations of the calculus. *Annals of Mathematics*, vol. 25.

Calhoun, M. E., Jucker, M., Martin, L. J., Thinakaren, G., Price, D. L., and Mouton, P. R. (1996). Comparative evaluation of synaptophysin-based methods for quantification of synapses. *Journal of Neurocytology* 25:821–828.

Calhoun, M. E., Kurth, D., Phinney, A. L., Long, J. M., Mouton, P. R., Ingram, D. K., and Jucker, M. (1998). Hippocampal neuron and synaptic Bouton number in aging C57bl/6 mice. *Neurobiology of Aging* 19:599–606.

Calhoun, M. E., and Mouton, P. R. (2001). New developments in neurostereology: Length measurement and 3d imagery. *Journal of Chemical Neuroanatomy* 21:257–265.

Cavalieri, B. (1635). *Geometria Indivisibilibus (Continuorum Nova Quadam Ratione Promota)*. Typis Clementis Ferronij. Bononi. Reprinted (1966) as *Geometria degli Indivisibili*. Unione Typographico-Editrice, Torino.

Chalkley, H. W. (1943). Method for the quantitative morphologic analysis of tissues. *Journal of the National Cancer Institute* 4:47–53.

Coggeshall, R., and Lekan, H. (1996). Methods for determining numbers of cells and synapses: A case for more uniform standards of review. *Journal of Comparative Neurology* 364:6–15.

Colonnier, M., and Beaulieu C. (1985). Counting of synaptic disks in the cerebral cortex. *Journal of Comparative Neurology* 231:175–179.

Cruz-Orive, L. M. (1976). Particle size-shape distributions: The general spheroid problem. *Journal of Microscopy* 107:233–253.

Cruz-Orive, L. M. (1976). Sampling designs for stereology. *Journal of Microscopy* 122:235–237.

Dam, A. M. (1979). Brain shrinkage during histological procedures. *Journal für Hirnforschung* 20:115–119.

DeHoff, R. T. (1962). The determination of the size distribution of ellipsoidal particles from measurements made on random plane sections. *Transactions Metallurgical Society AIME* 224:474–477.

DeHoff, R. T., and Rhines, F. N. (1961). Determination of number of particles per unit volume from measurements made on random plane sections: The general cylinder and the ellipsoid. *Transactions Metallurgical Society AIME* 221:975–982.

Delesse, M. A. (1847). Procédé mécanique pour déterminer la composition des roches. *Compte-Rendus d' Académie des Sciences, Paris* 25:544–545.

Dorph-Petersen, K. A., Nyengaard, J. R., and Gundersen, H. J. (2001). Tissue shrinkage and unbiased stereological estimation of particle number and size. *Journal of Microscopy* 204:232–246.

Duffy, K. B., Spangler, E. L., Devan, B. D., Guo, Z., Bowker, J. L., Janas, A. M., Hagepanos, A., Minor, R. K., DeCabo R., Mouton, P. R., Shukitt-Hale, B., Joseph, J. A., and Ingram, D. K. (2007). A blueberry-enriched diet provides cellular protection against oxidative stress and reduces a kainate-induced learning impairment in rats. *Neurobiology of Aging*, May 22.

Elias, H., Hennig, A., and Schwartz, D. (1971). Stereology: Applications to biological research. *Physiological Reviews* 51:158–200.

Fermi, L., and Bernardini, G. (1961). *La bilancetta (The Little Balance). Galileo and the Scientific Revolution*. Trans. Cyril Stanley Smith. Basic Books, New York.

Fox, C. H., Johnson, F. B., Whiting, J., and Roller, P. P. (1985). Formaldehyde fixation. *Journal of Histochemistry Cytochemistry* 33:845–853.

Francis, H. W., Rivas, A., Lehar, M., Saito, Y., Mouton, P. R., and Ryugo, D. K. (2006). Efficient quantification of afferent cochlear ultrastructure using design-based stereology. *Journal of Neuroscience Methods* 150:150–158.

Gardella D., Hatton, W. J., Rind, H. B., Rosen, G. D., and von Bartheld, C. S. (2003). Differential tissue shrinkage and compression in the z-axis: Implications for optical disector counting in vibratome-, plastic-, and cryosections. *Journal of Neuroscience Methods* 124:45–59.

Gardi, J. E., Nyengaard, J. R., and Gundersen, H. J. (2008). Automatic sampling for unbiased and efficient stereological estimation using the proportionator in biological studies. *Journal of Microscropy* 230:108–120.

Glagolev, A. A. (1933). On the geometrical methods of quantitative mineralogical analysis of rocks. *Transactions of the Institute of Economic Mineralogy (Moscow)* 59:1–47.

Gokhale, A. M. (1990). Unbiased estimation of curve length in 3–D using vertical slices. *Journal of Microscopy* 159:133–141.

Gokhale, A. M., Evans, R. A., Mackes, J. L., and Mouton, P. R. (2004). Design-based estimation of surface area in thick tissue sections of arbitrary orientation using virtual cycloids. *Journal of Microscopy* 216:25–31.

Goti, D., Katzen, S. M., Mez, J., Kurtis, N., Kiluk, J., Ben-Maiem, L., Jenkins, N. A., Copeland, N., Kakizuka, A., Sharp, A., Ross, C. A., Mouton, P. R., and Colomer, V. (2004). A mutant ataxin-3 putative-cleavage fragment in brain of Machado-Joseph disease patients and transgenic mice is cytotoxic above a critical concentration. *Journal of Neuroscience* 24:10266–102679.

Gundersen, H. J. G. (1977). Notes on the estimation of the numerical density of arbitrary profiles: The edge effect. *Journal of Microscopy* 111:219–223.

Gundersen, H. J. G. (1979). Estimation of tubule or cylinder LV, SV, and VV on thick sections. *Journal of Microscopy* 117:333–345.

Gundersen, H. J. G. (1986). Stereology of arbitrary particles: A review of number and size estimators and the presentation of some new ones, in memory of William R. Thompson. *Journal of Microscopy* 143:3–45.

Gundersen, H. J. G., Bagger, P., Bendtsen, T. F., Evans, S. M., Korbo, L., Nielsen, K., Nyengaard, J. R., Pakkenberg, B., Sørensen, F. B., Versterby, A., and West, M. J. (1988a). The new stereological tools: Disector, fractionator, nucleator and point sampled intercepts and their use in pathological research and diagnosis. *APMIS* 96:857–881.

Gundersen, H. J. G., Bendtsen, T. F., Korbo, L., Marcussen, N., Møller, A., Nielsen, K., Nyengaard, J. R., Pakkenberg, B., Sørensen, F. B., Versterby, A., and West, M. J. (1988b). Some new, simple and efficient stereological methods and their use in pathological research and diagnosis. *APMIS* 96:379–394.

Gundersen, H. J. G., Boysen, M., and Reith, A. (1981). Comparison of semiautomatic digitizer-tablet and simple point counting performance in morphometry. *Virchows Arch (Cell Pathol)* 37:317–325.

Gundersen, H. J. G., and Jensen, E. B. (1983). Particle size and their distributions estimated from line and point sampled intercepts, including graphical unfolding. *Journal of Microscopy* 131:291–310.

Gundersen, H. J. G., and Jensen, E. B. (1985). Stereological estimation of the volume weighted mean volume of arbitrary particles observed on random sections. *Journal of Microscopy* 138:127–142.

Gundersen H. J. G., and Jensen E. B. (1987). The efficiency of systematic sampling in stereology and its prediction. *Journal of Microscopy* 147:229–263.

Gundersen, H. J. G., Jensen, E. B. V., Kieu, K., and Nielsen, J. (1999). The efficiency of systematic sampling in stereology: Revisited. *Journal of Microscopy* 193:199–211.

Gundersen, H. J. G., and Osterby, R. (1981). Optimizing sampling efficiency of stereological studies in biology: Or "Do More Less Well!" *Journal of Microscopy* 121:63–73.

Haug, H., Kuhl, S., Mecke, E., Sass, N., and Wasner, K. (1984). The significance of morphometric procedures in the investigation of age changes in cytoarchitectonic structure of human brain. *Journal für Hirnforschung* 25:353–374.

Hazewinkel, M. (2002). Infinitesimal calculus. In *Encyclopedia of Mathematics*. Springer-Verlag, Berlin.

Hedreen, J. C. (1988). Lost caps in histological counting methods. *Anatomical Record* 250:366–372.

Hedreen, J. C. (1998). What was wrong with the Abercrombie and empirical cell counting methods? A review. *Anatomical Record* 250:373–380.

Hilliard, J. E. (1967). Determination of structural anisotropy. *Proceedings of the Second International Congress for Stereology.* Ed. H. Elias. Springer, Berlin.

Hilliard, J. E., and Lawson, L. R. (2003). *Stereology and Stochastic Geometry.* Computational Imaging and Vision Series, 1st ed. Springer, London.

Howard, C. V., and Reed, M. G. (2005). *Unbiased Stereology: Three-Dimensional Measurement in Microscopy (Advanced Methods),* 2nd ed. Taylor & Francis, London.

Howell, K., Hopkins, N., and McLoughlin, P. (2002). Combined confocal microscopy and stereology: A highly efficient and unbiased approach to quantitative structural measurement in tissues. *Experimental Physiology* 87:747–756.

Jensen, E. B., and Gundersen, H. J. G. (1993). The rotator. *Journal of Microscopy* 170:35–44.

Karlsson, L. M., and Gokhale, A. M. (1997). Stereological estimation of mean linear intercept length using the vertical sections and trisector methods. *Journal of Microscopy* 186:143–152.

Keiding, N., and Jensen, S. T. (1972). Maximum likelihood estimation of the size distribution of liver cell nuclei from the observed distribution in a plane section. *Biometrics* 28:813–829.

Kruman, I. I., Mouton, P. R., Emokpae, E., Cutler, R. G., and Mattson, M. P. (2005). Folate deficiency inhibits proliferation of adult hippocampal progenitors. *Neuroreport* 13:1055–1059.

Ladekarl, M., Boek-Hansen, T., Henrik-Nielsen, R., Mouritzen, C., Henriques, U., and Sørensen, F. B. (1995). Objective malignancy grading of squamous cell carcinoma of the lung: Stereologic estimates of mean nuclear size are of prognostic value, independent of clinical stage of disease. *Cancer* 76:797–802.

Larsen, J. O., Gundersen, H. J., and Nielsen, J. (1998). Global spatial sampling with isotropic virtual planes: Estimators of length density and total length in thick, arbitrarily orientated sections. *Journal of Microscopy* 191:238–248.

Lee, G. D., Aruna, J. H., Barrett, P. M., Lei, D.-L., Ingram, D. K., and Mouton, P. R. (2005). Stereological analysis of microvasculature parameters in a double transgenic model of Alzheimer's disease. *Brain Res Bull* 65:317–322.

Lei, D.-L., Long, J. M., Hengemihle, J., Ingram, D. K., O'Neill, J., Manaye, K. F., and Mouton, P. R. (2003). Effects of estrogen and raloxifene on neuroglia number and morphology in the hippocampus of aged female mice. *Neuroscience* 121:659–666.

Long, J. M., Ingram, D. K., Kalehua, A., and Mouton, P. R. (1998). Stereological estimation of microglia number in hippocampal formation of the mouse brain. *Journal of Neuroscience Methods* 84:101–108.

Long, J. M., Kalehua, A. N., Muth, N. J., Calhoun, M. E., Jucker, M., Hengemihle, J. M., Ingram, D. K., and Mouton, P. R. (1999). Stereological analysis of astrocyte and microglia in aging mouse hippocampus. *Neurobiology of Aging* 19:497–503.

Long, J. M., Mouton, P. R., Jucker, M., and Ingram, D. K. (1999). What counts in brain aging? Design-based stereological analysis of cell number. *Journal of Gerontology* 54A:B407–B417.

Loud, A. V. (1968). A quantitative stereological description of the ultrastructure of normal rat liver parenchymal cells. *Journal of Cell Biology* 37:27–46.

Luo, W., Chen, Y., Liu, A., Lim, H. M., Marshall-Neff, J., Black, J., Baldwin, W., Hruban, R. H., Stevensen, S. C., Mouton, P. R., Dardik, A., and Ballerman, B. F. (2004). Inhibition of accelerated graft arteriosclerosis by gene transfer of soluble FGF receptor-1 in rat aortic transplants. *Arteriosclerosis, Thrombosis, and Vascular Biology* 24:1081–1086.

Lyck, L., Santamaria, I. D., Pakkenberg, B., Chemnitz, J., Schrøder, H. D., Finsen, B., and Gundersen, H. J. (2009). An empirical analysis of the precision of estimating the numbers of neurons and glia in human neocortex using a fractionator-design with sub-sampling. *Journal of Neuroscience Methods* 182:143–156.

Manaye, K. F., Lei, D. L., Tizabi, Y., Davila-Garcia, M. I., Mouton, P. R., and Kelly, P. H. (2005). Selective cell loss in the hypothalamus of patients suffering from depression. *Journal of Neuropathology Experimental Neurology* 64:224–229.

Manaye, K. F., Turner, S., Xu, G., Lei, D.-L., Yukti, S., and Mouton, P. R. (2010). Alzheimer's type neuron loss in locus coeruleus of triple transgenic mice. *Journal of Alzheimer's Disease* 23, no. 4.

Manaye, K. F., Wang, P., O'Neil, J., Huafu, S., Tizabi, Y., Thompson, N., Ottinger, M. A., Ingram, D. K., and Mouton, P. R. (2007). Neuropathological quantification of dtg APP/PS1: Neuroimaging, stereology, and biochemistry. *AGE* 29:87–96.

Mandelbrot, B. B. (1977). *Form, Chance, and Dimension*. W. H. Freeman & Co., New York.

Marcario, J. K., Manaye, K. F., SantaCruz, K. S., Mouton, P. R., Berman, N. E. J., and Cheney, P. D. (2004). Severe subcortical degeneration in macaques infected with neurovirulent simian immunodeficiency virus. *Journal of Neurovirology* 10:1–13.

Maswood, N., Young, J., Tilmont, E., Zhang, Z., Gash, D. M., Gerhardt, G. A., Grondin, R., Yi, A., Roth, G. S., Lane, M. A., Mattison, J., Cohen, R. M., Mouton, P. R., Quigley, C., Mattson, M. P., and Ingram, D. K. (2004). Caloric restriction stimulates production of neurotrophic factors and attenuates neurochemical and behavioral deficits in a primate model of Parkinson disease. *Proceedings of the National Academy of Sciences of the United States of America* 101:18171–18176.

Matheron, G. (1972). Random set theory and applications to stereology. *Journal of Microscopy* 95:15–23.

Mathieu, O., Cruz-Orive, L. M., Hoppeler, H., and Weibel, E. R. (1981). Measuring error and sampling variation in stereology: Comparison of the efficiency of various methods for planar image analysis. *Journal of Microscopy* 121:75–88.

Mattfeldt, T., and Mall, G. (1984). Estimation of surface and length of anisotropic capillaries. *Journal of Microscopy* 153:301–313.

Miles, R. E. (1976). Precise and general conditions for the validity of a comprehensive set of stereology formulae. *Journal of Microscopy* 107:211–220.

Mogensen, O., Sørensen, F. B., Bichel, P., and Jakobsen, A. (1999). Mean nuclear volume: A supplementary prognostic factor in endometrial cancer. *International Journal of Gynecological Cancer* 9:72–79.

Morrison, J. M., and Hof, P. R. (1997). Life and death of neurons in the aging brain. *Science* 278:412–419.

Mouton, P. R. (2002). *Principles and Practices of Unbiased Stereology: An Introduction for Bioscientists.* Johns Hopkins University Press, Baltimore.

Mouton, P. R. (2005). *History of Modern Stereology.* International Brain Research Organization, http://www.ibro.info/.

Mouton, P. R. (2009). Applications of modern stereology to quantitative microscopy. In *Microscopy and Microanalysis*, vol. 15, ed. L. N. Brewer, S. McKernan, J. P. Shields, et al., 1526–1527. Cambridge University Press, London.

Mouton, P. R. (2011). Applications of unbiased stereology to neurodevelopmental toxicology. In *Developmental Neurotoxicology Research: Principles, Models, Techniques, Strategies, and Mechanisms*, ed. C. Wang and W. Slikker. John Wiley & Sons, Hoboken, NJ.

Mouton, P. R., Chachich, M. E., Quigley, C., Spangler, E., and Ingram, D. K. (2009). Caloric restriction attenuates cortical amyloidosis in a double transgenic mouse model of Alzheimer's disease. *Neuroscience Letters* 464:184–187.

Mouton, P. R., Durgavich, J., and Ingram, D. K. (2005). Automatic estimation of size parameters using Verified Computerized Stereoanalysis. *Image Analysis & Stereology* 24:1–9.

Mouton, P. R., Gokhale, A. M., Ward, N. L., and West, M. J. (2002). Stereological length estimation using spherical probes. *Journal of Microscopy* 206:54–64.

Mouton, P. R., and Gordon, M. (2010). Design-based stereology and video densitometry for assessment of neurotoxicological damage. In *Neurotoxicology.*

Target Organ Toxicology Series, 3rd ed., ed. G. J. Harry and H. A. Tilson, 243–267. Informa Healthcare USA, New York.

Mouton, P. R., Kelley-Bell, B., Tweedie, D., Spangler, E. L., Perez, E., Carlson, O. D., Short, R. G., Decabo, R., Chang, J., Ingram, D. K., Li, Y., and Greig, N. H. (2010). Effects of age and lipopolysaccharide (LPS)-mediated peripheral inflammation on numbers of central catecholaminergic neuron. *Neurobiology of Aging*, Epub online, November.

Mouton, P. R., Long, J. M., Lei D.-L., Howard, V., Jucker, M., Calhoun, M. E., and Ingram, D. K. (2002). Age and gender effects on microglia and astrocyte number in brains of mice. *Brain Research* 956:30–35.

Mouton, P. R., Martin, L. J., Calhoun, M. E., Dal Forno, G., and Price, D. L. (1998). Cognitive decline strongly correlates with cortical atrophy in Alzheimer's dementia. *Neurobiology of Aging* 19:371–377.

Mouton, P. R., Pakkenberg, B., Gundersen, H. J. G., and Price, D. L. (1994). Absolute number and size of pigmented locus coeruleus neurons in the brains of young and aged individuals. *Journal of Chemical Neuroanatomy* 7:185–190.

Mouton, P. R., Price, D. L., and Walker, L. C. (1997). Empirical assessment of total synapse number in primate neocortex. *Journal Neuroscience Methods* 75:119–126.

Oaklander, A. L., Stocks, E. A., and Mouton, P. R. (2003). Number of Langerhans' immune cells in painful and non-painful human skin after shingles. *Archives of Dermatological Research* 294:529–535.

Ohm, T. G., Busch, C., and Bohl, J. (1997). Unbiased estimation of neuronal numbers in the human nucleus coeruleus during aging. *Neurobiology of Aging* 18:393–399.

O'Neil, J. N., Mouton, P. R., Tizabi, Y., Ottinger, M. A., Lei, D.-L., Ingram, D. K., and Manaye, K. F. (2007). Catecholaminergic neuron number in locus coeruleus of aged female dtg APP/PS1 mice. *Journal of Chemical Neuroanatomy* 34:102–107.

Pakkenberg, B., Andersen, B. B., Jensen, G. B., Korbo, L., Mouton, P. R., Møller, A., Regeur, L., and Øster, S. (1992). The impact of the new stereology on the neurosciences—neurostereology. *Acta Stereologica* 11:157–164.

Pakkenberg, B., and Gundersen, H. J. G. (1997). Neocortical neuron number in humans: Effect of sex and age. *Journal of Comparative Neurology* 384: 312–320.

Pakkenberg, B., Pelvig, D., Marner, L., Bundgaard, M. J., Gundersen, H. J., Nyengaard, J. R., and Regeur, L. (2003). Aging and the human neocortex. *Expermimental Gerontology* 38:95–99.

Paumgartner, D., Losa, G., and Weibel, E. R. (1981). Resolution effect on the stereological estimation of surface and volume and its interception in terms of fractal dimension. *Journal of Microscopy* 121:51–63.

Perry, T. A., Holloway, H. W., Weerasuriya, A., Mouton, P. R., Duffy, K., and Mattison, J. A. (2007). Evidence of GLP-1–mediated neuroprotection in an animal model of pyridoxine-induced peripheral sensory neuropathy. *Experimental Neurology* 203:293–301.

Perry, T. A., Weerasuriya, A., Mouton, P. R., Holloway, H. W., and Greig, N. H. (2004). Pyridoxine-induced toxicity in rats: A stereological quantification of the sensory neuropathy. *Experimental Neurology* 190:133–144.

Ripley, B. D. (1976). The foundations of stochastic geometry. *Annals of Probability* 4:995–998.

Roberts, J., and Goldberg, P. B. (1976). Some aspects of the central nervous system of the rat during aging. *Experimental Aging Research* 2:531–542.

Rosival, A. (1898). Uber Geometrische Gesteinsanalysen. *Verh der K. K. Geolog. Reichanstalt (Wein)*, pp. 143–175.

Saper, C. B. (1996). Any way you cut it: A new journal policy for the use of unbiased counting methods. *Journal of Comparative Neurology* 354:5.

Smith, C. S., and Guttman, L. (1953). Measurement of internal boundaries in three-dimensional structures by random sectioning. *Trans AIME* 197:81–111.

Sørensen, F. B. (1989). Stereological estimation of nuclear volume in benign melanocytic lesions and cutaneous malignant melanomas. *American Journal of Dermatopathology* 11:517–520.

Sterio, D. C. (1984). The unbiased estimation of number and sizes of arbitrary particles using the disector. *Journal of Microscopy* 134:127–136.

Stocks, E. A., McArthur, J. M., Griffin, J. W., and Mouton, P. R. (1996). An unbiased method for estimation of total epidermal nerve fiber length. *Journal of Neurocytology* 25:11–18.

Subbiah, P., Mouton P. R., Fedor, H., McArthur, J., and Glass, J. D. (1996). Stereological analysis of cerebral atrophy in human immunodeficiency virus-associated dementia. *Journal of Neuropathology and Experimental Neurology* 55:1032–1037.

Tandrup, T. (1997). The optical rotator. *Journal of Microscopy* 186:108–120.

Thomson, E. (1930). Quantitative microscopic analysis. *Journal of Geology* 38: 193–222.

Van Vré, E., van Beusekom, H., Vrints, C., Bosmans, J., Bult, H., and Van der Giessen, W. (2007). Stereology: A simplified and more time-efficient method than planimetry for the quantitative analysis of vascular structures in different models of intimal thickening. *Cardiovascular Pathology* 16:43–50.

Vijayanshankar N., and Brody, H. (1977). Aging in the human brain stem: A study of the nucleus of the trochlear nerve. *Acta Anatomica* 99:169–172.

Vijayanshankar, N., and Brody, H. (1979). A quantitative study of the pigmented neurons in the nuclei locus coeruleus and subcoeruleus in man as related to aging. *Journal of Neuropathology and Experimental Neurology* 38:490–497.

Vitruvius, M. P. (ca. first century BC). *De Architectura Libri Decem* (*The Ten Books on Architecture*).

Weibel, E. R. (1979). *Stereological Methods*, vol. 1. *Practical Methods for Biological Morphometry*. Academic Press, London.

Weibel, E. R. (1980). *Stereological Methods*, vol. 2. *Theoretical Foundations*. Academic Press, London.

Weibel, E. R., and Gomez, D. G. (1962). A principle for counting tissue structures on random sections. *Journal of Applied Physiology* 17:343–348.

Weibel, E. R., Stäubli, W., Gnägi, H. R., and Hess, F. A. (1969). Correlated morphometric and biochemical studies on the liver cell. I. Morphometric model, stereologic methods, and normal morphometric data for rat liver. *Journal of Cell Biology* 42:68–91.

West, M. J. (1999). Stereological methods for estimating the total number of neurons and synapses: Issues of precision and bias. *Trends in Neurosciences* 22:51–61.

West, M. J., Coleman, P. D., Flood, D. G., and Troncoso, J. C. (1994). Differences in the pattern of hippocampal neuronal loss in normal ageing and Alzheimer's disease. *Lancet* 344:769–772.

West, M. J., Slomianka, L., and Gundersen, H. J. G. (1991). Unbiased stereological estimation of the total number of neurons in the subdivisions of the rat hippocampus using the optical fractionator. *Anatomical Record* 231: 482–497.

Wicksell, S. D. 1925. The corpuscle problem: A mathematical study of a biometric problem. *Biometrika* 17:84–99.

INDEX

The letter *t* following a page number denotes a table. The letter *f* following a page number denotes a figure. The word *defined* following a page number denotes a glossary term.

and, 135; equation for total N, 83; fractionator and, 85; frame area and, 77f; guard zone and, 157; height, 75, 85; mean object volume and, 102; optical, 78, 80, 81f, 82f, 86f, 158; optimal sampling parameters and, 108; pair, 76f, 81f; physical, 76, 76f, 158; principle, 75, 76, 76f, 78, 79, 85, 87, 90, 102; probability and, 106, 109; rare events and, 90, 91, 93, 94; size distribution and, 96, 104; D. C. Sterio and, xi; tsf and, 160; validation and, 79, 87

distributions, 52, 74, 94 107, 108, 138, 156 (defined); size and, 74f, 96, 104, 104f; spatial, 22

"Do More, Less Well" (Ewald Weibel), 14, 42, 105, 109f, 115, 116, 120, 146, 154, 156 (defined)

edge effect, 77f

efficiency, 13, 39, 105, 120, 156 (defined); automatic data management and, 140; axis for vertical section and, 54, 55f; biological variability and, 154; computerized stereology systems and, 147; nucleator vs. rotator, 103f; point counting and, 16; point- vs. pixel counting, 37, 38f, 40, 42; probability and, 109; probes for length estimation and, 60; rare event protocol and, 90, 91, 93–95; sampling, 12, 85, 154

Elias, Hans, ix, 1–4

electron microscopy, *Stereologer* and, 141

embedding, 156 (defined); paraffin, 30, 32, 148

error analysis, 156 (defined)

error variance (CE^2), 117, 156 (defined); contributions to, 14; partitioning error and, 120, 146

estimate(s), 156 (defined); biased, 79, 152; local size, 100, 103; measurements and, 34–35, 38, 39; rare events and, 94; unbiased, 34, 35, 62, 66, 132, 148

estimator(s): geometric probes and, 21, 22; length and length density, 63, 64; number and number density, 84, 84t, 146; number-weighted, 96, 146;

probability and, 23; ratio (density), 63; 88f; surface area and surface density, 58, 58f, 62, 69, 69f; volume (local), 96, 97f, 103; volume (regional), 26; volume fraction and, 20; volume-weighted, 96, 98, 99f

Euclid, 5–8

Euclidean geometry, 156 (defined)

exclusion lines, 156 (defined); planes and, 76, 80, 84f; unbiased counting frame and, 160

expected (true) value, ix, 35, 108

expert-based morphometry, 2, 147

first-order stereological parameters, ix, 116, 120, 148, 156, 157 (defined); anisotropy and, 49; fractal dimensions and, 69; geometric probes and, 22, 110; probability and, 105

fixation, 30, 123, 128, 157 (defined); cyroprotection and, 124; tissue shrinkage and, 88f, 122

Floderus (correction factor), 74t

Ford Motor Company, 142

fractal dimension, 68, 68f; surface area and length, 67, 68

fractionator, 34, 85–87, 86f, 157 (defined); optical, 109, 118f, 158; peer review and, 150t; rare event protocol and, 90–92, 94

geometric probes, 22, 45, 71; isotropy and, 46, 56; needle problem and, 46; number- vs. volume-weighted, 96–100, 102, 144, 158, 160; probability of intersection and, 50, 105, 106, 108; virtual, 133

geometry: classical (Euclidean), 100, 156 (defined); biased stereology and, ix, 35; computerized systems and, 147; correction factors and, 148; probability theory and, 64, 101; stochastic (stocastic), 148, 154, 159

Gokhale, Arun, xii, 57f, 58f, 148

Great Britain (fractal coastline), 66, 67

guard volume (zone), 78f, 82f, 157 (defined); optical disector and, 80

Gundersen, Hans, J. G., xi, 148; counting rules and, 76, 78, 80

numerical aperture (NA), 81f, 94, 129, 130, 135
numerical density (ratio estimator for N_A or N_V), 73, 83–85, 83f; vs. total number, 72, 73f

object(s), man-made vs. naturally occurring, 5, 6f
optical disector, 78f, 79, 80, 81f, 82f, 158 (defined); counts of brain cells (neurons) and, 89f; fractionator and, 85, 86, 86f
optical fractionator, 85, 87, 109, 158 (defined); counts of brain cells, 118f
optical scanning, thin focal plane, 81, 82, 135; anisotropy and, 45
optimization for maximum efficiency, 158 (defined)
orientation, anisotropic, 45, 50, 50f, 51; cycloid line and, 55
orientation bias, 158 (defined)
overprojection (Holmes effect), 130, 131f, 158 (defined)

Pakkenberg, Bente, xi
paraffin-embedding, 158 (defined)
parameter dimensions, 22, 23, 23f, 48, 62; fractals and, 68, 69; second-order (second-moment), 33, 37f; stereological, first-order, 49, 157 (defined)
pattern recognition, 158 (defined)
physical disector, 76, 76f, 79, 83f, 158 (defined)
pixel counting, vs. point counting, 37–40, 42
pilot study, importance, 115, 150, 151, 152; cell counts, 118
point grid, 25; random placement, 71
point sampling intercepts, 99f; and cancer staging, 97–99, 99f
precision, vs. accuracy, 33, 35, 36f, 39; bias and, 35; measurement vs. estimation, 37, 38, 42; probability and, 109; rare events and, 90; *Stereologer* and, 141
probability, 42, 44, 105–7; equal sampling and, 108
profile counting, vs. object counting, 72, 72f, 73, 73f, 75, 75f

Querschnitt (German "cross section"), 83; ΣQ (sum disappearing cross sections), 83, 85, 92, 93

random sampling, systematic, 69f, 143, 159 (defined); biological variability and, 133; Cavalieri volume and, 31; mean cell volume and, 97, 102; point grid and, 99, 138; section sampling and, 28, 91, 111, 136, 139, 144, 154; surface area exercise, 70
rare event protocol, 90, 91, 112; efficiency compared to stepping motors, 94, 112; equation to estimate total N, 87, 93; example, 95
ratio estimator, age-related brain changes, 88; conversion to absolute (total) parameter, 63, 64, 84
recognition bias, 159 (defined)
reference space, 144, 159 (defined); bookending and, 124; object distributions and, 107; rare event protocol and, 94; reference volume calculation and, 30; vs. region of interest, 12; sampling and, 83, 95, 106, 111, 112, 138
reference trap, 20, 84, 88f, 145, 159 (defined); brain aging and, 87; as source of bias, 21, 64, 83, 88f; uniform tissue shrinkage and, 63
region of interest (ROI), 159 (defined); compared to reference space, 12; sectioning and, 48
rotator, 103, 103f, 104, 104f, 150t, 159 (defined); mean cell volume, 145
Royal Academy of Sciences (Paris), Buffon and, 43; Delesse and, 16

sample estimate, unbiased, 33; error variance and, 156; precision and, 36, 158
sample size (power), 118, 154; effect size and, 112; pilot study and, 115; sampling stringency and, 117, 119, 146
sampling, 12–14, 33, 34, 36f, 145, 154; biased vs. unbiased, 148; CE and, 146; fractionator, 86f; hierarchy and, 143; high and low precision and, 120;

sampling *(continued)*
    optimization for maximal efficiency,
    stringency, 113, 115, 116; *Stereologer*
    and, 138; stringency, 113; interval, 13,
    154; systematic-random, 143, 144;
    uniform-random, 160
sampling error, 12, 14, 154, 159 (defined);
    between-section, 14, 111, 117, 155;
    optimization for maximal efficiency,
    114, 116; pilot study and, 156, 158;
    sampling stringency and, 118f; within-
    section, 76, 117, 160
scale dependence (fractal dimensions),
    64, 66f, 68
second-order stereological parameter(s),
    ix, 22, 140, 137f, 159 (defined);
    design-based stereology and, 156;
    Stereologer and, 135
section sampling fraction (SSF), 12, 86f,
    91–93, 144, 159 (defined)
section thickness, 32, 31; cut (block
    advance), 79, 124; fractionator and,
    85, 86f, 92; postprocessing (final), 30;
    Stereologer and, 103
shape bias, 159 (defined)
shrinkage, 20, 21, 83, 85, 88; differential
    and, 30, 32, 63, 84; fixation and, 122
size bias, 48; Cavalieri principle and, 27;
    Corpuscle Problem and, 31, 73; disector
    principle and, x, xi, 75, 78; number-
    weighted parameter and, 158; unbiased
    counting frame and, 78
size estimators, local, 96, 97f
software for computerized stereology,
    130, 134, 141; complete systems and,
    140; high-throughput systems and,
    146; *Stereologer* and, 77f, 78, 135,
    138, 159
space balls (isotropic sphere probes), 59f,
    60, 60f, 145, 160 (defined); magnifi-
    cation and, 135; section thickness
    and, 124
spatial distribution, ix, 22
Stereologer, 78, 134, 140, 141, 146,
    159; hardware requirements, 135,
    130; software, 138
stereology: assumption- and model-
    based, 148–49; coined, ix, xii; com-
    puterized systems, 133–37, 141, 146;

design-based, 8, 156 (defined); ISS
    and, 4; organic, 47, 110, 126, 140,
    143, 148, 150, 151; unbiased, 149
Sterio, D. C., xi, 75, 79, 86; disector
    and, 78
stochastic geometry, 47, 154, 159 (defined);
    probability theory and, 64, 101, 148
surface area, ix, 22, 49, 57f, 62, 70,
    157; anisotropy and, 50, 51, 55f;
    Buffon and, 44; coastline of Great
    Britain and, 67; comparison two
    studies, rat liver, 67; cycloids and, 54;
    dimensions, 22, 23f; fractal dimension
    and, 65, 68, 69; IUR sections and, 53;
    needle problem and, 45, 69; vertical
    sections and, 55; virtual cycloids and,
    129, 145, 158, 160
survey (analogy to stereology), 142
systematic error (bias), 36, 79, 116, 120;
    unbiased stereology and, 149
systematic-random-sampling, 125, 144,
    145, 154, 160 (defined); fractionator
    and, 91; *Stereologer* and, 133, 136

thickness sampling fraction (TSF), 85,
    92, 93, 160 (defined)
thin sections, Cavalieri volume estima-
    tion and, 12; area estimation, 53, 56,
    69; length estimation, 57f, 62, 124,
    135; point-sampled intercepts and,
    96, 97, 130f; profile counting and, x
tissue processing, x, 31, 93, 120, 153,
    154; Cavalieri method for volume
    estimation and, 30; section thickness,
    32; uncertainty and, 120; reference
    trap and, 20, 84

unbiased counting frame, 75, 160
    (defined); inclusion and exclusion
    lines and, 156, 157; optical disector
    and, 80, 158
unbiased methods, 35, 36f, 37
uncertainty (nonstereological bias), 120,
    121, 131, 146, 153; optical artifacts
    and, 130; reference spaces and, 12
underprojection, 160

variation (error), 105, 114, 118f, 140,
    150, 156, 160 (defined); assumption-